ALSO BY JEROME GROOPMAN

The Measure of Our Days

Second Opinions

The Anatomy of Hope

The Anatomy of Hope

HOW PEOPLE
PREVAIL IN THE FACE
OF ILLNESS

Jerome
Groopman, M.D.

RANDOM HOUSE / NEW YORK

RANDOM HOUSE and colophon are registered trademarks of Random House, Inc.

Library of Congress Cataloging-in-Publication Data
Groopman, Jerome E.
The anatomy of hope : how people prevail in the face of illness / Jerome Groopman.
p.   cm.
Includes bibliographical references and index.
ISBN 0-375-50638-1
1. Patients—Psychology.   2. Sick—Psychology.   3. Hope—Health aspects.
4. Healing—Psychological aspects.   5. Mental healing.
6. Physician and patient.   I. Title.
R726.5.G76   2004   616'.001'9—dc21   2003046692

Printed in the United States of America on acid-free paper
Random House website address: www.atrandom.com
2 4 6 8 9 7 5 3 1
First Edition

Book design by J. K. Lambert

For

STEVEN, MICHAEL, AND EMILY.

What greater hope do we have

than our children?

Pandora, the first mortal woman, received from Zeus a box that she was forbidden to open. The box contained all human blessings and all human curses. Temptation overcame restraint, and Pandora opened it. In a moment, all the curses were released into the world, and all the blessings escaped and were lost—except one: hope. Without hope, mortals could not endure.

Contents

The Anatomy of Hope

W hy do some people find hope despite facing severe ill-ness, while others do not? And can hope actually change the course of a malady, helping patients to prevail?

I looked for the answers in the lives of several extraordinary patients I cared for over the past thirty years. They led me on a journey of discovery from a point where hope was absent to a place where it could not be lost. Along the way, I learned the dif-ference between true hope and false hope, and describe times when I foolishly thought the latter was justified. There were also instances when patients asserted their right to hope and I wrongly believed they had no reason to do so. Because they held on to hope even when I could not, they survived. And one woman of deep faith showed me that even when there is no longer hope for the body, there is always hope for the soul. Each person helped me see another dimension of the anatomy of hope.

Hope is one of our central emotions, but we are often at a loss when asked to define it. Many of us confuse hope with optimism, a prevailing attitude that "things turn out for the best." But hope differs from optimism. Hope does not arise from being told to "think positively," or from hearing an overly rosy forecast. Hope, unlike optimism, is rooted in unalloyed reality. Although there is no uniform definition of hope, I found one that seemed to capture what my patients had taught me. Hope is the elevating feeling we experience when we see—in the mind's eye—a path to a better future. Hope acknowledges the significant obstacles and deep pitfalls along that path. True hope has no room for delusion.

Clear-eyed, hope gives us the courage to confront our circumstances and the capacity to surmount them. For all my patients, hope, true hope, has proved as important as any medication I might prescribe or any procedure I might perform. Only well into my career did I come to realize this. During my training in medical school classrooms and on hospital teaching rounds, we saw patients as fascinating puzzles. Making a diagnosis and finding the optimal therapy were essentially detective work. We mined the stories of patients' lives for clues. Family background, experiences at the workplace, travel, personal habits, and relationships all gave hints to solving the clinical mystery. The family history provided information about inherited genes and how they predisposed people to one disorder or another; the workplace suggested potential exposure to carcinogenic chemicals or poisonous metals; travel could bring contact with arcane pathogens that populate far regions of the world; habits like smoking and drinking could promote pathology; and relationships helped uncover sexually transmitted diseases like syphilis, HIV, and gonorrhea.

Solving a complex case and identifying the best treatment is indeed an exhilarating intellectual exercise. But the background and stories of patients' lives give doctors the opportunity to probe another mystery: How do hope, and despair, factor into the equation of healing?

For nearly three decades I have practiced hematology and oncology, caring for patients with cancer, blood diseases, HIV, and hepatitis C. I have also labored in my laboratory, studying the genes and proteins that these disorders derange. During much of that time, at the bedside and at the laboratory bench, I failed to consider the impact of hope on my patients' illness. Yes, I gave the customary nod to it, but then I would focus squarely on interpreting their laboratory reports, reading their CAT scans, and studying their biopsies—all essential to diagnosis and treatment, but incomplete. What was missing had to be learned from experience. I had to be tested—not on paper but by overcoming adversity, both as physician and as patient.

A vast popular literature exists contending that positive emotions affect the body in health and disease. Much of it is vague, unsubstantiated, merely wishful thinking. These books depict hope as a magic wand in a fairy tale that will, by itself, miraculously restore a patient. As a rational scientist, trained to decode the sequence of DNA and decipher the function of proteins, I fled the fairy-tale claims of hope. In effect, I slammed the door on hope and closed off my mind to seriously considering it as a catalyst in the crucible of cure.

Personal experience opened my mind. For some nineteen years after failed spine surgery, I lived in a labyrinth of relapsing pain and debility. Then, through a series of chance circumstances, I found an exit. I felt I had been given back my life. I recognized that only hope could have made my recovery possible. Rekindled hope gave me the courage to embark on an arduous and contrarian treatment program, and the resilience to endure it. Without hope, I would have been locked forever in that prison of pain. But I also sensed that hope had done more than push me to take a chance and not give up. It seemed to exert potent and palpable effects not only on my psychology but on my physiology.

As a scientist, I distrusted my own experience, and set out on a personal journey to discover whether the energizing feeling of hope can in fact contribute to recovery. I found that there is an authentic biology of hope. But how far does it reach? And what are its limits? Researchers are learning that a change in mind-set has the power to alter neurochemistry. Belief and expectation— the key elements of hope—can block pain by releasing the brain's endorphins and enkephalins, mimicking the effects of morphine. In some cases, hope can also have important effects on fundamental physiological processes like respiration, circulation, and motor function. During the course of an illness, then, hope can be imagined as a domino effect, a chain reaction in which

each link makes improvement more likely. It changes us profoundly in spirit and in body.

Every day I look for hope, for my patients, for my loved ones,
and for myself. It is an ongoing search. Here I tell what I have
found.

While all of the incidents in this book are true, the names and personal characteristics of the individuals involved have been changed in order to protect their privacy. Any resulting resemblance to other persons, living or dead, is entirely coincidental and unintentional.

The Anatomy of Hope

CHAPTER 1

Unprepared

In July 1975, I entered my fourth and final year of medical school at Columbia University in New York City. I had completed all my required courses except surgery and was eager to engage in its drama.

Surgeons acted boldly and decisively. They achieved cures, opening an intestinal blockage, repairing a torn artery, draining a deep abscess, and made the patient whole again. Their art required extraordinary precision and self-control, a discipline of body and mind that was most evident in the operating room, because even minor mistakes—too much pressure on a scalpel, too little tension on a suture, too deep probing of a tissue— could spell disaster. In the hospital, surgeons were viewed as the emperors of the clinical staff, their every command obeyed. We students were their foot soldiers. I was intoxicated with the idea of being part of their world.

The surgical team I joined was headed by Dr. William Foster. Foster was a tall, imposing man with sharp features like cut timber. His rounds began at dawn, followed by two or three surgeries that lasted until late afternoon. As is typical in a teaching hospital, all of Dr. Foster's patients were assigned to medical students who learned the basics of diagnosis and treatment by following cases. Not long after I began the course, I was designated as the student to help care for Esther Weinberg, a young woman who had a mass in her left breast.

Esther Weinberg was twenty-nine years old, full-bodied, with almond-brown eyes. She was a member of the Orthodox Jewish community in Washington Heights, the neighborhood adjoining Columbia's medical school. When I entered her room, Esther was lying on the bed, reading from a small prayer book. Her head was covered by a blue kerchief in the typical sign of modesty among married Orthodox women, whose hair, as a manifestation of their beauty, is not to be seen by men other than their husbands.

"I'm Jerry Groopman, Dr. Foster's student," I said by way of introduction. I wore the uniform of the medical student, a short, starched white jacket with my name on a badge over the right breast pocket. The badge conspicuously lacked the initials "M.D." Esther quickly took my measure, her eyes lingering over my name badge.

I did not reach out to shake her hand. Men do not touch strictly Orthodox women, even in a casual way.

Esther's eyes returned to my name badge, then to my face. I guessed at what was crossing her mind: whether my not shaking her hand indicated that I was Jewish and knowledgeable of the Orthodox prohibition, or simply an impolite student. "Groopman" was Dutch in origin, not a giveaway. Dr. Foster had described Esther as anxious, and I felt that disclosing our shared heritage would put her at ease.

"*Shalom aleichem,*" I said, the traditional greeting of "Peace be with you."

Instead of offering a welcoming smile, her face drew tight.

Following protocol, I began the clinical interview, which includes taking a family and social history. Esther Weinberg, née Siegman, was born in Europe in 1946. Her family was from Leipzig, Germany, and of its more than one hundred members, only her parents had survived the Nazi camps. The Siegmans immigrated to America in the early 1950s. Esther married at the age of nineteen, had her first child—a girl—a year after the wedding, and then twin girls eighteen months later. Her father died of a stroke not long after. Over the last year, she had worked as the personal secretary for the owner of a cleaning service in midtown Manhattan; her job was strictly clerical, without exposure to toxic solvents that can be carcinogenic.

One of the primary risk factors for breast cancer is a family history of the disease. Esther had limited knowledge of those who had perished in the war, but she recalled no afflicted relatives. Another major risk is prolonged and uninterrupted exposure to estrogen, which occurs when menarche, the onset of menses, happens at a very young age, or when pregnancy occurs later in life or not at all. But Esther had entered puberty at thirteen, a typical time, and carried and nursed three children in her twenties. This early motherhood would, if anything, lower her risk for breast cancer.

I conducted the physical examination that I was taught to perform specifically on women, to convey a sense of propriety and respect for their body. I covered each breast in turn as I palpated for irregularities. I was taken aback by what I found. The mass in her left breast was very large, about the size of a golf ball, easily felt above the nipple. There were many lymph nodes in the left armpit, also large and rock-hard.

For a cancer to grow to this size, and to spread into the adjoining lymph nodes, takes many months, if not years. Its prognosis, dictated by the dimensions of the tumor and the numbers of lymph nodes containing metastatic deposits, was very poor. How could a seemingly attentive young woman have waited so long to consult a doctor?

I did not ask. Dr. Foster strictly defined boundaries for students on his surgical team. Our role was to observe and learn, to do only what he told us to do.

"We will be making rounds with Dr. Foster later in the day," I said. "I wish you the best with the surgery."

"God willing" was her reply.

I started to leave.

Esther called after me, "Can I talk to you?"

"Of course," I said. A patient choosing to talk to us students made us feel very much like the doctors we wanted to be.

"Maybe later," she said uncertainly.

That afternoon, William Foster stood at the foot of Esther Weinberg's bed, flanked on his left by his three students, and on his right by the team's two residents. The waning July daylight cast long shadows across the room. I summarized the reason for admission, the physical findings, and the planned procedure, directing my words to Dr. Foster. The mass was almost certainly malignant, and it appeared to be quite advanced; it would first be treated by surgery, followed by chemotherapy. I went on with my charge as a student, reviewing for the team what Mrs. Weinberg had been told by Dr. Foster in his office about the impending operation. After she was anesthetized in the operating room, a biopsy would be taken of the mass, and if it proved to be malignant, as expected, a radical mastectomy would be performed right away. This was the approach handed down from William Halsted, an eminent surgeon who practiced in the early 1900s at Johns Hopkins.

Dr. Foster nodded and walked deliberately to the left side of the bed. He held Esther Weinberg's hand in his. He asked if she had any questions about the impending operation.

"Will Dr. Groopman be with me when I wake up after the surgery? I'd like him there."

Dr. Foster shot me a brief, quizzical look.

I was unsure why Esther wanted me at her side when she regained consciousness. I studied her face for a clue, but it revealed none.

"Mr. Groopman, like every student, follows his patients from the time of admission into the operating room and then through postoperative care. Be assured that I will discuss fully with you what we found at surgery and what next steps need to be taken."

———

Esther Weinberg's case was the first on the day's schedule. I scrubbed next to Dr. Foster and the senior resident. There was no idle chatter before surgery. We marched single file into the OR, Dr. Foster leading, the senior resident behind him, and I last, befitting my status. The anesthesiologist had already put Esther under. Foster nodded to me, and I swabbed an iodinelike antiseptic in concentric circles over the skin of her left chest. Then I laid sterile drapes around the painted breast.

Since beginning the surgery course on the first of the month, I had assisted in several operations and seen how the operative field was treated, as if it were a domain distinct from a larger living human being. The surgeon initially identified the relevant anatomical landmarks, like a surveyor delineating his planes. This promoted psychic detachment, lowering the emotional temperature and facilitating the intense concentration the cutting required. A stylized sequence reinforced this mind-set. Each set of incisions was followed by a formal appraisal of the newly exposed anatomy and a resetting of landmarks. The aim was to fully encompass the diseased region with minimum destruction to surrounding healthy tissues and maximum preservation of normal structures. But today's operation was different. In the event of a radical mastectomy, total destruction of the normal anatomy was planned. The mammary tissues of the breast would be removed, along with the muscles overlying the chest wall, including the pectoralis and all the lymph nodes of the armpit. What would remain were scar and ribs. This draconian approach was rooted in Halsted's contention that cancer cells migrated stepwise from the primary tumor into the surrounding tissues and then, much later, through the bloodstream to distant sites like liver and bone. Only by extirpating a complete block of flesh on the chest could the surgeon remove the cancer cells hiding beneath the breast. Dr. Foster had lectured at length on how Halsted's insight had advanced the treatment of breast cancer from a plethora of haphazard operations to a uniform and highly scientific surgery.

Dr. Foster delineated the margins of the breast mass above Esther's left nipple and then instructed the resident to biopsy it. He made an incision and retrieved a wedge of gritty, glistening tissue. A pathologist was called to perform a "frozen section." He would flash-freeze part of the mass and immediately examine it under the microscope to determine whether malignant cells were present. If he saw them, the mastectomy would proceed.

Our wait in the OR was a short and silent one. Dr. Foster seemed deeply absorbed in his thoughts, and neither the resident nor I dared disturb him. The pathologist reentered the OR. His face was grave. There was no doubt about what he had seen on the frozen section.

Dr. Foster began making bold strokes around the circumference of the breast. I held a cautery, and as Dr. Foster cut, he directed me to burn the ends of small bleeding vessels. Wisps of acrid smoke with the distinctive odor of charred flesh wafted from the cauterized vessels. My stomach tightened.

After over three hours, the dissection was complete. When the breast en bloc was lifted from the chest wall, globules of fat and lymph seeped from its base, the underlying muscles raw and bleeding. More than a dozen lymph nodes had been removed from the axilla.

My mind drifted. I looked up from the operative site to the tube in Esther's mouth that delivered the anesthetic. She would awake to a drastic change in her form. It was impossible to predict how she would react. But I imagined that being Orthodox and married, with an established faith and family, would help her cope.

The recovery room was a large open space brightly illuminated by overhead panels of fluorescent lights. At the entryway was a board on which each patient's name was written next to the number of the bed the patient occupied. A nurse and I wheeled Mrs. Weinberg on a gurney into the recovery room, which was filled with other patients. We stopped to write her name on the board. "Foster's radical?" the clerk at the board asked. I nodded. "Bed six," he directed.

Using the undersheet for leverage, we lifted Esther onto her assigned bed. A harsh chorus of voices resonated in the room—"Run more saline in that line"; "His pressures are low, ramp up the dopamine"; "Check her oxygen, she looks a little blue around the lips." Dr. Foster marched through the din like a reigning royal. He dispatched the resident to scrub for the next case. Ordinarily, I would have returned to the wards to assist in chores. But Esther had been promised that I would be at her side when she awoke.

I looked down at her. Beads of perspiration glistened on her forehead. Her hair, matted by the sweat, was cropped close to be easily covered by a *tichel*.

Dr. Foster squeezed her right hand and said her name several times until her eyes opened and stayed open. She struggled to focus, the effect of the anesthetics still in her system.

"Mrs. Weinberg, the surgery is completed." Dr. Foster paused. "I'm sorry, we had to remove the breast."

Esther was silent for a while, then nodded slowly and turned toward me. "Groopman, you'll understand," she whispered. I held her gaze for a long moment, but her glazed eyes seemed opaque. She soon drifted back to sleep.

Just outside the recovery room, Dr. Foster stopped abruptly. "What did she mean by that?"

I had no idea.

———

Three days later, I received some clue to Esther's enigmatic comment.

On morning rounds Dr. Foster told Esther that because of the large size of the tumor and its spread to more than a dozen lymph nodes, the cancer was likely to recur soon. Chemotherapy would now be given to destroy the lurking cells in her body; it would begin once the mastectomy healed. "Chemotherapy is unpleasant," Dr. Foster allowed, "but potentially lifesaving." Foster himself would oversee the drug treatment, a common practice among cancer surgeons of his generation.

I visited Esther alone later that day. I had decided to take the initiative and ask her to explain her statement in the recovery room. But it didn't prove necessary.

"You should call me Esther. May I call you Jerry?"

"Of course," I said.

"Jerry, I was unsure at first if I could trust you. Can I trust you?"

Out of reflex, I extended my hand and grasped hers. Esther did not withdraw. She smiled, but it was a smile forced to fight back tears.

"My cancer is a punishment from God," Esther said flatly.

I began to reply, but Esther stopped me. "Wait."

She explained that Markus Weinberg, her husband, was from the same German-Jewish community in Washington Heights and had been chosen for her by her parents. He was twenty-one at the time, and she was nineteen. They had met twice before the wedding, and she knew on each occasion that she could not love him. She described him as meek and complacent, with little interest in the larger world. He worked as a grocer in a family business not far from their home. Esther exchanged few words with him, mostly about who would do what chore, or whether one of the children was progressing well in her schoolwork.

"I felt I could breathe only out of the house. Inside, I couldn't."

She knew that even with the formless skirts extending to her ankles, the billowing blouses with sleeves reaching her wrists, and her hair covered by the *tichel,* men turned when they passed her on the street. And she saw that her boss, middle-aged with a family in New Jersey, paid special attention to her. He told jokes that seemed designed to make her laugh, complimented her work and small changes in her appearance, like when she wore amber earrings that he said offset her eyes.

Esther said she knew what her boss wanted, and although she didn't believe for a minute that he loved her, to be desired was enough for her to feel that she was not suffocating.

"He closes the blinds in his office, and he takes me on the floor." Tears streamed over the crest of her cheeks.

I handed her a box of tissues, but she waved it away. "I should be crying for shame," she said. "But I'm not. I'm crying because I know he won't want me anymore. He'll find me repulsive, at best someone to pity. There'll be another girl. And the little that I had in life that was my own will be gone."

Esther looked at me with pleading eyes. Was I supposed to affirm her sin and acknowledge its punishment? Or was I to sympathize with her plight and, after hearing her confession, absolve her?

I was not ordained to do either. So I released her hand and silently left the room.

———

I felt off balance under the weight of her secret. Each day after rounds, I was alone with her. Esther did not offer her hand in greeting, and I did not extend mine. Uncertain what, if anything, I should say, I kept the conversation strictly clinical: Was the chest pain decreasing? Did she have feverish feelings or sweats? Had the swelling in her left arm and fingers changed?

On the day of Esther's discharge, Dr. Foster stood at her bedside and uncovered her chest. The suture lines were tight, and there was no sign of infection. There was the expected edema due to the removal of lymph channels that normally drain the arm. He pointed out to the residents and students how quickly a radical mastectomy begins to heal in a young woman. He declared the surgery to be a success.

"Now, my office has you scheduled for a return appointment in two weeks," Dr. Foster said to Esther. "If anything changes in the interim, if you have unusual pain, if fever develops, or the arm becomes suddenly red or more swollen, be sure to contact me before then."

Esther nodded.

"Once there has been sufficient healing, we will start on the chemotherapy. Mr. Groopman will be back later to make sure everything is in order for you to go home."

Shortly after four P.M., I knocked on Esther's door. A male voice said, "Please, come in." Standing by Esther's bedside was a compact man in a loose-fitting gray suit with a black fedora.

"Markus Weinberg. I am Esther's husband," he said.

I introduced myself and, smiling, said that everything was ready for her discharge.

Esther's face betrayed no emotion. I reiterated the doses of pain medication she should take when needed, and the date of her follow-up appointment with Dr. Foster.

"Thank you for your attention to my wife," Mr. Weinberg said. "Perhaps—perhaps we could invite you for a Sabbath meal? We are not far, maybe fifteen blocks from the medical center."

Esther's face remained masklike.

I did not know how to respond, unsure whether Esther had suggested or agreed to this plan, and uncertain whether accepting it would violate what Dr. Foster viewed as the proper boundaries with patients.

"Thank you, but it may be a bit much right now for Mrs. Weinberg, after the surgery."

"No, Esther's mother, Gertrude, lives not far from us, two blocks away, and she is preparing food for us, helping with the children, until Esther is through all this."

I again said I appreciated the invitation, but my schedule was unpredictable in the coming months.

Markus Weinberg looked at his wife.

"We would be honored to have you," Esther said.

On the following Friday, shortly before sunset, I threaded my way north from the medical center through Washington Heights. It had rained steadily through the day, and the heat of the August sun lifted patches of vapor from the glistening asphalt. Esther's seeming ambivalence about the invitation amplified mine. Knowing her secret made me feel complicit in her other life. I strangely feared that Markus Weinberg or Esther's mother would somehow read on my face what I knew.

One part of me wanted to step back, but another propelled

me forward. I wanted to prove to myself that I was already more than a student, able to act as a doctor. A physician, I told myself, who crossed the boundary from the professional to the personal with a patient and family would have the poise and skill to control the situation.

The Weinbergs lived on the second floor of a squat yellow brick building several blocks before the George Washington Bridge. Markus Weinberg was waiting for me on the street.

The Weinbergs' apartment was a small, tidy space, with a long corridor that led to a central dining area. Esther stood as we entered. She was dressed in a starched white blouse and a blue skirt, with small diamond earrings that caught the light from the flickering Sabbath candles. Markus moved to the head of the table and indicated that I was next to him, facing Esther. Two of their girls sat at the far end on either side of Gertrude, a heavyset woman with sharp eyes and a commanding manner. The girls had Esther's auburn hair.

A soft quiet descended. Markus Weinberg blessed his daughters. Then he turned toward Esther and sang the traditional song of a husband to his wife on Sabbath eve:

"Ayshes chayil me y'imtza . . ."

A woman of valor, who can find?
She is precious far beyond rubies.
Her husband trusts in her, and he shall lack nothing thereby.
She renders him good and not evil all the days of her life.

She opens her hand to the needy and extends her hand
  to the poor.
She is robed in strength and dignity and cheerfully faces the
  future.
She opens her mouth with wisdom; her tongue is guided
  by kindness.
She tends to the affairs of her household and eats not the
  bread of idleness.
Her children come forward and bless her; her husband, too,
  and praises her:
"Many women have done superbly, but you surpass them all."
Charm is deceitful and beauty is vain, but a God-revering
  woman is much to be praised.
Give her honor for the fruit of her hands; whenever people
  gather, her deeds speak her praise.

Esther winced during the verses.

"Did you take the pain medicine?" her mother asked.

"I did. Maybe I'll take another after the meal."

Markus quickly said the blessings over the wine and bread. Esther insisted on helping serve, and not surprisingly, she favored her right hand. Her left fingers were swollen like small sausages. It was still too early to know how much of the edema would recede or whether the mastectomy's disruption of lymph channels from the limb to the chest would make the hand swelling a permanent condition.

"Are you a doctor?" Esther's eldest daugher, Rebecca, asked. I calculated from Esther's admission interview that she would be around nine years old.

"Not yet."

"How much longer do you have to study?"

I explained that there were four years of medical school, and that when I completed this last year, I would begin as an intern, then serve as a resident. Following this, I planned to specialize and do research on blood diseases and cancer.

"What's cancer?" Rebecca asked.

"Rebecca, our guest came to eat," Gertrude said sternly. *"Frayg nicht kein shilas."*

Ask no questions.

The drum of the window air conditioner filled a long silence.

I looked down the length of the table. Platters were filled with roasted chicken and brisket of beef, and casseroles held potato and noodle kugel. I speared a piece of chicken with the serving fork and placed it on my plate. The clink of the metal against the china resonated through the room.

"It may pass ninety again tomorrow," Markus Weinberg said. "It was ninety-two today."

"And very humid," I added.

Esther picked distractedly at her plate. Despite Markus's prompting, she did not take dessert.

The meal ended with Markus Weinberg leading grace. His daughters seemed to take special pleasure in the singing of thanksgiving for God's bounty.

"They have beautiful voices," I commented.

"Like angels," Esther said. For a fleeting moment, her lips trembled. She seemed swept away by the force of her feelings and looked down. When her gaze returned, she was composed again.

Markus Weinberg stood and thanked me for coming. He instructed his daughters to retire to their room and ready themselves for bed. I prepared to leave, complimenting Esther's mother again on the food.

"I need a breath of air," Esther said. "I'll walk the doctor a few blocks."

A hint of uncertainty passed over Markus's face, but he did not object.

The day's heat still hung over the street. A group of young men sat on a stoop at the end of the block, drinking bottles of soda. They eyed Esther as she passed.

"You see my life," Esther said. "Everything is hidden. Even what's wrong with me can't be spoken."

I replied that there were different opinions about how much to tell children when a parent fell ill. "They don't even know that I was in the hospital," Esther said. "It was as if I disappeared and then returned. I carry it all myself, alone. Markus talks to the rabbi, my mother talks to God. I can't talk to either."

"Perhaps you should talk confidentially to the rabbi," I offered.

Esther looked hard at me. "Are you serious? In the eyes of the community, I am a *zonah*, a whore," she said. Then, "I am not going to take the chemotherapy."

I stared in disbelief. "You have to do what Dr. Foster said. He told you what was found at surgery. You have a very high risk of the cancer recurring, and that would be fatal. Esther, the chemotherapy gives you a chance to live."

"Good Sabbath," Esther said, and turned back to her apartment.

I sat at my desk in the dormitory, the taste of the rich Sabbath meal lingering on my tongue. I was on call through the weekend, and scheduled to admit a patient with a blocked pancreatic duct. My surgery textbook lay open to the chapter on the anatomy of the pancreas and the different disorders that obstruct it. Dr. Foster would quiz me about the material on Monday-morning rounds. But I could not disengage from Esther Weinberg.

It was clear now why Esther had not sought medical attention until long after the cancer appeared. Also clear was her cryptic remark about how I would understand.

The Bible often equates disease with punishment for sin. When Miriam, the sister of Moses, slanders his wife, God makes her leprous. While the rabbis later interpreted her leprosy as a metaphor to signify that a person is consumed by envy and eaten away by jealousy, in the minds of fundamentalists such literary glosses hold little weight. In their eyes, God shows two faces to man. One is the God of mercy, long forbearing and abundant in kindness, always ready to forgive his imperfect creatures when they repent: Miriam regrets her transgression, and her leprosy is healed. The other God is one of justice, stern and demanding, setting forth codes of conduct that, when violated in the extreme, trigger harsh retribution.

By not seeking care at the first sign of the cancer, Esther had started on a sure path to death. She had wavered from that path by consulting Dr. Foster. If she received aggressive treatment right away, she had at least a chance of living. Her refusal was a return to the road of self-destruction.

I had said she could trust me, had given her my word I would guard her secret. Yet I wished she had never confided in me. I had been disturbed by the knowledge but also, I admitted to myself, flattered that she chose to put her trust in me. Somehow I thought that our shared heritage would make me able to help her. I now saw how shallow and naive that idea was. I had stepped out of the foundation of fact and knowledge into a quagmire of emotion and imagination. I was in over my head.

I was a fourth-year student, not a doctor. My responsibilities for patients were limited to what Dr. Foster said I could and should do. And, as a student, when I was uncertain about a diagnosis or did not fully comprehend a choice of therapy, I would never hesitate to ask a senior resident or the attending physician. Whom should I ask for guidance now?

There were clergy at the hospital who specialized in assisting patients, but Esther had declined to meet with the medical center's rabbi during her hospitalization. Psychotherapists were regularly brought in on complex cases, but the Orthodox community harbored a deep suspicion of followers of Freud and those who prescribed medications that modulated the mind.

Should I be held to my promise? I decided I would keep Esther's secret for now. Her follow-up appointment with Dr.

Foster was in two weeks. I was sure that in William Foster's commanding presence, Esther would comply and go ahead with the chemotherapy.

I returned to the anatomy of the pancreas.

But Esther did not begin her treatment as planned. I was in the clinic when she called and told the secretary she didn't feel up to coming in and starting the drugs. Later in the week, I overheard the clinic secretary informing Dr. Foster that Esther had canceled the rescheduled appointment.

My stomach tightened when Dr. Foster mentioned that Esther was resisting her treatment. He had called her personally to find out why, and they had spoken at length.

"She sees her cancer as some sort of divine judgment," Foster tersely stated. I said she had told me this, too, not knowing what more Esther had revealed to Dr. Foster. His tone did not suggest further discussion. He seemed to be exerting the same steely self-control that I had observed in the OR. In parting, he said he would continue to meet with Esther until reason prevailed.

I finished the course three months later with a guilty sense of relief. It took that long for Dr. Foster to convince Esther. Her first chemotherapy treatment coincided with my last days on surgery.

Medical students have relatively short-lived interactions with patients, the relationship constrained by the weeks they work in a particular clinic or ward. I spent the ensuing months abroad, taking elective courses in England and Israel. After graduation, I moved to Boston for my internship. Only later did I learn from classmates who stayed at Columbia that Esther had sought medical attention too late. The cancer reappeared, first in her bones, then in her liver, and finally in her lungs. She died at the age of thirty-four.

Often I am tempted to refrain from telling Esther's story, or from telling it in full. Looking back at my actions reveals painful ignorance.

For my initial two years of medical school, I had sat in lecture halls absorbing a wealth of information about anatomy, physiology, pharmacology, and pathology. During the last two years, I had stood rapt at the bedside, taking in the words of master clinicians who revealed the subtleties of physical examination and the fine points of medical treatment. Brimming with new knowledge, I thought I was fully ready to assume the care of people. I mistook information for insight. While I was well prepared for the science, I was pitifully unprepared for the soul.

I never learned why Esther reversed her decision. While attending physicians like Dr. Foster instructed us about the manifestations of diseases and showed us the operative techniques to remedy them, the subjects of hope and despair were not part of our curriculum. Conversations like the one between Dr. Foster and Esther occurred behind closed doors. Students and interns and residents were not privy to the words that a doctor used to change a patient's mind and heart. Deeply uncomfortable with what had transpired, I was reluctant to ask Dr. Foster about Esther that last week of the course, and he did not deem it necessary to tell me.

For several months, Esther's case haunted me, but then it largely faded from my mind. Internship was an intensely focused and heady time, and the culture at the Massachusetts General Hospital was one of brash independence and sure action. Acceptance into the training program in internal medicine was highly selective, with some six hundred applicants for eighteen places. Its interns and residents, like those at other teaching hospitals, went on to become leaders of academic medicine. To hone leadership qualities, the hospital kept a philosophy of sink-or-swim. While we worked during the day as a team composed of two interns and one senior resident, each intern was alone on call at night. Although you could contact the senior resident for backup, the unspoken rule was that this showed weakness, either in knowledge or in character. After a long, sleepless night on the wards or in the emergency room, you were supposed to present yourself on morning rounds to your fellow intern and resident as an "iron man" who had held the fort singlehandedly.

What counted in front of your peers and the supervising attending physicians was making astute diagnoses and intervening with effective treatments. This boiled down to finding clues in the patient's clinical history, discerning some telling abnormality in the physical examination, interpreting correctly the shadows on the X ray, and knowing the appropriate dose of a drug or proper placement of an instrument to treat the disorder effectively.

There was a fast drumbeat to those days and nights, and I thrived on the speed and the stress. The system was meant to make you efficient, and among the interns and residents, efficiency was joined to impatience. Only what was clinically essential was worth addressing. Although lip service was paid to the patient's emotional state, it was largely ignored. If anyone had talked about the psyche and the soul, that would have brought raised eyebrows from all of us young doctors who saw our future as steeped in hard science. We would be the leaders of laboratories where DNA was decoded and the structure of proteins delineated, the chairmen of clinical departments where new instruments were designed and experimental medicines tested. The idealism that had brought many of us into medicine was recast as corny. Issues of soul and spirit were best left to those who did not aim for the academic stratosphere, thus triaged to social workers and hospital clergy.

There were moments, though, when I was reminded of Esther. Patients would not take their medicine or would fail to show up for an X ray. When they next came into the clinic, I would repeat my instructions and reiterate the necessity for the treatment or procedure, but I could see in their eyes a lack of light, in their face a cold indifference. I had little time, though—and, I had to admit to myself, little motivation—to probe deeply into why they seemed to be in despair, unable to take the necessary steps

toward better health. It was only years later that I began to realize how closely such patients were connected to Esther.

At first I saw Esther's case in a narrow light. Her fundamen-
talistic theology equated disease with punishment for sin. She
believed she was undeserving of hope because of her transgression. God had made his judgment, and God's omnipotence and
omniscience meant there was no reason for her to have hope,
no reason to persevere. Why endure when everything about her
present and future was irrevocably determined? It was a chilling
and awful state of mind, a living hell with no possibility of exit.

Later, I began to see that this was too limited a view. Esther's
religious background and beliefs were the particular language,
the metaphors, in which hopelessness was cast.

Hope can arrive only when you recognize that there are real
options and that you have genuine choices. Hope can flourish
only when you believe that what you do can make a difference,
that your actions can bring a future different from the present. To
have hope, then, is to acquire a belief in your ability to have some
control over your circumstances. You are no longer entirely at
the mercy of forces outside yourself.

At the time, I had no insight into any of this, and my blindness
was sustained for many, many years. During my internship and
residency, the patients like Esther who resisted effective treatment, mortgaging their future due to hopelessness, remained
opaque to me. On team rounds, we rationalized such cases with
fault-finding rubrics like "recalcitrant" and "noncompliant." These
labels caused us to distance ourselves further from the patients.

Labels shifted the responsibility from our shoulders to theirs and, if anything, solidified the sense of despair that surrounded them.

Esther had no hope, and it seemed to me a miracle that Dr. Foster had convinced her to take the chemotherapy. I had been confused and perplexed by her intimacy with me, and by her despair, but even more confounded by my inability to reach her. It was part of my youthful hubris that I wanted to be the one responsible for her improvement—but I didn't know what it took to give her hope. Dr. Foster did know, it appears, but such lessons weren't part of the curriculum. I wish I had learned what Dr. Foster told her and how he was able to finally break through. He would have shown me, beyond scalpels and sutures, the ways a doctor tries to make a person whole again. But Dr. Foster never would have shared with a student what he said to change her mind; nor, at the time, would I have understood the deeper importance of that event. That lesson would begin several years later.

CHAPTER 2

False Hope, True Hope

M y residency in Boston ended in 1978, and I moved to Los
    Angeles to begin a specialty fellowship in blood diseases
and cancer medicine at UCLA. I was twenty-seven years old,
recently engaged, with little money to support a future family.
To supplement my fellowship stipend, I worked as a "moon-
lighter," covering the practice of Dr. Richard Keyes in a rural
town two hours north of L.A. that I'll call Russell.

Keyes was in his fifties, a broad-shouldered man with thick gray
hair, wire-rimmed spectacles, and an accent that revealed his New
England origins. He had long, delicate fingers; after college he
had spent two years at a conservatory in Austria studying piano
before he decided on a career in medicine. Keyes had trained in
oncology at Stanford and could have stayed there and ascended
the academic ladder, but he chose to relocate to Russell; it was
where he thought he could do the most good. Since there were

no other cancer specialists in the area, Keyes hired young doctors like me to take his calls and visit his hospitalized patients from Saturday night through Monday morning. Friday afternoons, when I was not on call at UCLA, I would drive north on the interstate, past the mansions on Mulholland Drive, through the smoggy San Fernando Valley, and then some miles beyond Simi to Russell. Large and small farms dotted the surrounding countryside. Irrigation lines sprayed fine mists into the arid air, creating hazy rainbows around migrant workers who picked lettuce and cauliflower.

Keyes's office was in a three-story stucco medical office building with an orange tile Spanish roof. One particularly hot day in August, he introduced me to a patient named Frances Walker. She was a slim African-American woman, fifty-two years old, with close-cropped graying hair and a full face that framed serious eyes. Frances was accompanied by her teenage daughter, Sharon, very much a younger version of her mother.

I had read Frances's chart in preparation for the consultation. She worked as a cashier in a local dry-goods store. After World War II, she and her husband had come to California from Missouri in search of work. He had died years before of kidney disease, and Sharon was their only child. Frances had had no prior medical problems other than high blood pressure. Some three weeks earlier, traces of blood had been found in her stool during a yearly physical examination. A colonoscopy showed a tumor in the lower bowel. A surgeon resected the tumor but found that the cancer had spread to multiple lymph nodes and invaded the left lobe of her liver. Frances had Stage D colon cancer, the most advanced stage. At today's visit, Keyes would explain to her the operative findings and the rationale for a course of treatment. I would attend to her over the weekend.

Richard Keyes greeted the Walkers warmly. He drew a curtain that divided the consultation room in half and asked Frances to step behind it, undress, and put on a hospital gown. While she did, he chatted with Sharon. "Your mother told me you're a junior this year at Russell Regional?"

"I am."

Dr. Keyes smiled broadly. "Thinking about colleges yet?"

"My grades are good, and I scored high on the SATs. So my adviser says to aim for an Ivy League school."

"Definitely consider my alma mater," Keyes said. "Of course, when I was there—shortly after the Stone Age—it was all men, all white, and all beer parties, all the time. Dartmouth isn't that way anymore—and it's a better place."

Richard Keyes drew open the curtain. Frances lay on the examining table.

"So your daughter is a star at school," he said.

"She's a blessing to me," Frances replied.

Keyes asked about her symptoms since the surgery. Frances said her appetite had returned quickly, she had no fevers or chills, and her bowel movements had resumed in a normal fashion.

"All good signs," he said. Keyes covered Frances's lower pelvis with a white sheet, then lifted her gown to expose her abdomen. He ran his fingers along the fresh scar and palpated deeply under her right rib cage, seeking the edge of her liver. "Any pain?"

"No, not really."

He nodded and said that the healing had gone well. "Why don't you get dressed and then we'll talk?"

He drew the curtain closed, sat down at the desk, took out a clean sheet of paper, and began to make notes. I sat beside him, across from Frances and Sharon when they joined us. As he finished his note, he glanced at me briefly and then addressed his patient.

"Frances, all of the cancer was removed from your bowel and the surrounding lymph nodes. A few small spots of tumor were found on the left side of the liver. But we have chemotherapy to help take care of them."

A wave of relief passed over Frances's face. Sharon exhaled audibly.

"The chemotherapy I will give you is very active against those spots in the liver. I expect some side effects, like mouth sores, diarrhea, and anemia, but you'll be monitored very closely. All of the side effects can be managed and will ultimately reverse.

"Dr. Groopman will see you over the weekend after today's treatment. The chemotherapy continues until we shrink the remaining cancer down. Any questions?"

Frances thought for a moment and then said no, she understood what needed to be done.

Keyes turned to Sharon. "Don't hesitate to call me if you seriously consider Dartmouth."

"I won't."

We both stood as the Walkers left the room.

Keyes completed his clinic note and handed it to me to review. The note ended with the stylized phrase that commonly concludes consultations: "Patient and family understand the risks and benefits of the proposed therapy."

I had paid close attention to how Richard Keyes communicated with the Walkers. He knew as well as anyone that Stage D colon cancer was very rarely, if ever, curable. "Cure" meant total eradication of the cancer so it would never recur. Chemotherapy, in cases like Frances's, was palliative, meaning it was given in an attempt to slow the growth of the malignant cells in the liver. Palliation, then, was a trade-off, the price of the months of treatment hopefully offset by the time gained. The price was unquestionably high: The toxic chemotherapy would cause the complications Keyes had enumerated. The gain could be one or even two years of productive life. But it was also quite possible that no extension of life would occur, despite the treatment. Frances might suffer months of chemotherapy for nothing.

I was still feeling my way on how to communicate a poor prognosis to patients and their families. Not once during my schooling, internship, or residency had I been instructed in the skill. Despite the plethora of modern technologies—CAT scans and angiography and electron microscopy—learning how to care for patients was still very much like being an apprentice in a medieval guild. You closely and repeatedly observed master craftsmen at their work and then, largely on your own, tried your hand at it. Richard Keyes had been practicing oncology for some two decades. I was sure he had chosen his words deliberately. In my own dealings with the Walkers, I wanted to be sure that I had correctly observed, and would respect, the boundaries that he had set.

"When I'm with the Walkers this weekend, if direct questions come up, I should emphasize remission, correct?" I said to Richard.

"Yes. I certainly wouldn't look at Frances and say: 'Madam, the cancer in your liver will kill you.'" Keyes let his reply linger. "What's the point of that? All it does is make the remaining time even more miserable. Or cause her to panic and refuse palliation. And what good would it do to tell Sharon? She's the apple of Frances's eye. A bright, motivated girl like that—I bet she's the first one in her family who ever considered going to college, and she's shooting for the Ivies. Her grades as a junior are critical. She'd go into a tailspin."

Keyes looked at his watch. "We have two more people to see before you drive back to L.A.," he said. "Each doctor has his own style, his own way of doing things. Jerry, believe me, for patients in situations like this, too much information is overwhelming."

I did not press Richard further. Medicine was not metaphysics. Truth was never absolute. I would follow his lead.

As expected, Frances had some nausea and dry heaves over the weekend, but these were tempered with antiemetic medication. She seemed in good spirits despite the side effects. "I'm a fighter," she said. "Have been all my life. I passed through all sorts of difficulties. With God's help and you doctors, I'll pass through this one as well."

Sharon looked at her mother with a mixture of pride and worry. I was heartened by Frances's words, knowing that a strong spirit was needed to endure the months of toxic treatment that lay ahead.

One on-call weekend, I had to admit Frances to the hospital. Although she pushed herself to drink fluids like broths rich with salts and nutrients, the painful mouth ulcers from the chemotherapy had made it impossible to take in enough, and she became dehydrated. I administered intravenous saline and gave her topical pain medication for the ulcers. Richard Keyes returned on Monday, and by Wednesday Frances was discharged home. Another weekend Frances had spiking fevers and lancinating abdominal cramps and diarrhea. I again brought her into the hospital, gave her antibiotics, and in a few days' time, her symptoms subsided. Frances endured the treatment and its side effects without once complaining. Sharon adopted an unflinching attitude as well.

"Look at that CAT scan!" Richard Keyes exclaimed to the Walkers three months into the therapy. "This is the liver," he said, pointing to a large gray form. He took his pen and traced the borders of several small black circles within the organ. "Those are the deposits we are treating. They're about half the size of when we started."

"Does that mean," Frances asked hesitantly, "I am partly cured?"

Keyes smiled warmly. "You are well on the way to a remission."

"Thank God. It's going away."

Sharon closed her eyes and bowed her head in a silent prayer.

In the field of cancer medicine, there are standardized objective criteria to assess the response of a tumor to a particular treatment. The term "remission" refers to the degree of shrinkage in the size of a cancer. A "partial remission" signifies at least a 50 percent decrease in a tumor's vertical and horizontal diameters. A "complete remission" denotes that the cancer can no longer be detected even by sophisticated techniques like CAT scan. But a complete remission does not necessarily signify a "cure." Cure, in medical parlance, means not only the complete disappearance of all detectable cancer in the body but that it will never return.

Richard Keyes again had spoken the literal truth, that sufficient shrinkage had occurred in Frances's liver metastases to signify a remission. It was a partial one, since the diameters of the tumors had decreased by over half. With further therapy, the partial remission might become complete. I, too, rejoiced that the cancer was in retreat. Frances might turn out to be one of the fortunate patients who gained more time, and quality time, from palliative chemotherapy. And, I admitted to myself, not explaining the difference between remission and cure to the Walkers had made the visit easier.

Nineteen seventy-nine was my first winter in Southern California, and aside from occasional rain, the weather seemed to take no notice. Each morning I jogged around the perimeter of UCLA, nearly five miles, and if I had time in the evenings, I swam outdoors in the university's fifty-meter pool. Pam, my fiancée, was two years behind me in her medical training and had stayed in Boston to complete her residency at Massachusetts General Hospital. I felt that I was the explorer, sent ahead to survey this new land. On Sundays I would jog to the Pacific and look at the horizon, filled with the sense that this was the far edge of America, a place to define oneself anew. Working closely with Richard Keyes in Russell was also part of that redefinition, I told myself.

One Friday in early January, I was scheduled to cover Keyes's practice. The waiting room was full when I arrived, and Frances was among the patients. When I shook her hand, I noticed that it trembled. Sharon's smile could not mask her concern. I escorted both women into the examining room. Frances slowly undressed.

Richard Keyes entered, greeted the Walkers warmly, and began to palpate gently the quadrants of Frances's abdomen. He asked her to take a deep breath and then pressed under her right ribs. He stopped midway when Frances winced. "That hurts," she said.

He repeated the maneuver, and again Frances winced.

I looked at a recent laboratory printout clipped to her chart, and saw that her liver-function tests were elevated as they had not been before.

"Your liver edge is tender, and your blood tests are slightly abnormal," Keyes said. "Sometimes the chemotherapy can inflame the liver as a side effect."

That was a possibility, but a remote one. The more likely

explanation was that the cancer had become resistant to the treatment and the tumors were growing. The elevated liver-function tests were the footprints of malignant cells marching through the organ.

"You are due for a follow-up CAT scan in a week," Keyes said. "Let's see each other in two weeks' time to discuss the results. Until then I'll give you a prescription for some pain medication. Don't be reluctant to use it if you need to."

Frances said she would take the pills when the pain came on, thanked him, and left with Sharon.

Keyes took off his reading glasses and rubbed his eyes. "You know, it really doesn't make a difference clinically if it is the cancer and not the chemo. There's little we can do about it. By telling Frances and Sharon now, we just add another few weeks of worry. This way they have something to cling to for a little longer." Richard paused and looked at me kindly. "You're at the beginning of your career, Jerry. In my experience—and I am talking many years in this community with these kinds of people, who are absolutely the salt of the earth, honest, hardworking, religious, devoted to family—sustained ignorance is a form of bliss."

What Richard said made sense to me. It seemed that focusing on relapse would just cause anguish for Frances and Sharon.

Keyes added, "Maybe she'll be lucky and it will turn out to be a side effect from the drugs."

Two weeks later, when I returned to Russell, I saw Frances's name on the patient list. The first thing I did was search for the report of her CAT scan. The liver metastases had more than doubled in size, and new deposits had appeared in the spleen. The organs looked as though they had been riddled by large-caliber bullets, the kind that expand upon impact, leaving gaping holes. The scan also showed that fluid was building up in the abdomen, a condition termed "ascites." This complication was particularly difficult to manage clinically. Some chemotherapy drugs could accumulate in the retained fluid and not be cleared from the body, thereby prolonging their toxic effects. The ascitic fluid could also surround and choke the stomach and bowel, causing pain and preventing the absorption of vital nutrients.

Even though I had limited experience as an oncology fellow, I knew that patients like Frances rarely survived over a few months. Still, I was shocked and saddened by how rapidly the cancer was growing.

"Dr. Keyes called," his nurse informed me. I put aside the CAT scan. "He said he can't break away from a case in the hospital. He asked that you begin with Frances. Once he's free, he'll join you."

I entered the consultation room. Sharon sat in the corner, clutching a textbook. Frances was already gowned and lying on the examining table. As I greeted her, I noticed a faint tinge of yellow in her eyes. It was jaundice, an indication that the cancer was blocking the liver's excretion of bile: an ominous sign.

"How are you?" I asked.

Frances shook her head. "Very tired. I have no appetite. I have to force myself to eat, since the food doesn't go down easily."

I found what I expected on physical examination. Her abdomen was so distended from the ascites that it pressed her navel outward like a bubble.

I told the Walkers that ascites had developed, and explained that Frances's difficulty with ingesting food was due to pressure on her stomach from the fluid.

"We need to drain the ascites to relieve the pressure. It's a relatively simple procedure. We put a catheter in at the midline below your navel. You should feel better afterward."

"Is the fluid from the cancer?" Sharon asked.

I said it was. I told Frances and Sharon about the CAT scan.

"Then that means it's spreading quickly, doesn't it?" Sharon asked sharply. The edge to her voice was unmistakable.

"Sharon . . ." Frances said with a tone of parental authority.

I nodded in answer to Sharon's question. She screwed her face into a tense knot.

"My appetite is so poor," Frances said. "I have no energy. I felt for a while that something was wrong. . . . But Dr. Keyes said it was from the chemotherapy treatments."

Sharon's eyes blazed. "I thought you and Dr. Keyes said that the chemotherapy could cure her."

"He didn't—we didn't—quite say that," I replied. "We said that there was a good chance of going into remission, which happened." I spoke in a deliberate, even voice, and explained what remission meant and how it differed from cure. Frances's eyes became moist. I handed her a box of tissues. She looked away as she dabbed her eyes.

"Why didn't you tell us before?" Sharon asked.

I momentarily cast my eyes down. "Colon cancer behaves this way," I said, "shrinking for a while from the treatment, then becoming resistant to it and growing again. I am sorry."

I called the nurse and made arrangements for the ascites to be drained at the hospital. The procedure needed to be done under radiological guidance so the inserted catheter would not inadvertently lacerate the liver or spleen.

"You'll feel better very soon, when the fluid is removed," I told Frances. She acknowledged my words and slowly lifted herself from the examining table. I closed the curtain. Sharon helped her to dress.

Richard Keyes arrived nearly an hour later. "Too many balls in the air," he said by way of explanation for his absence.

Over the ensuing weeks, Richard was always late when Frances was scheduled. When he arrived, he informed me that he had seen an urgent case in the ER or assisted another doctor with a procedure. By the time he arrived, Frances and Sharon were either gone or there was little time for conversation.

The last time I saw the Walkers was in early March. Frances was unable to eat more than a few bites of solid food. If drinks were too cold or too hot, she regurgitated them. Each drainage of the ascites provided only a few days of relief before the fluid reaccumulated. Richard was on vacation back east for a week. I was spending my vacation time covering his practice.

Frances declined further chemotherapy after hearing my frank recitation of the data on its chance of working. She found sitting, eating, even talking effortful. I said we should focus on relieving pain, then I shifted to address matters of emotional comfort.

"You mentioned you are close with your minister."

"I am."

"Does he visit at home now that it's difficult for you to go to church?"

"Yes," she said.

"Often?"

"Yes."

The anguish was palpable in her every word. Sharon did not speak to me.

It was time to make arrangements for hospice care. Frances simply nodded at my suggestion and said she had already discussed it with her minister.

Frances was taken in a wheelchair by a nurse to the laboratory for a set of blood tests. Sharon lingered behind. For a long moment I couldn't meet her gaze.

"I know Dr. Keyes tried his best," I said. "He'll make sure when he returns that your mom is comfortable."

Sharon pondered my words. "I guess he didn't think people like us are smart enough, or strong enough, to handle the truth."

"It wasn't a question of smart enough," I said. "And it wasn't specific to your mom and you. Dr. Keyes and I were trying to spare you the worry."

"Well, you were both wrong."

As I drove back to Los Angeles, Sharon's words echoed in my

THE ANATOMY OF HOPE

mind. A sense of shame and guilt gripped me. Richard and I had failed the Walkers. It had been a delusion to tell myself that what Richard had done, and what I had embraced as his apprentice, was "for the best" for them. Ignorance was not bliss, not when it mattered. By abandoning the truth, Richard and I had abandoned Frances, and through our deception we left Sharon alienated and bitter.

Frances's last weeks were spent at home, hospice nurses attending to her along with Sharon. Richard was the one who informed me of her death.

"She did not suffer at the end," Keyes commented. "The hospice team did a wonderful job."

I nodded in agreement, thinking how he and I had not.

"The nurse told me that Sharon will be living with cousins. It's a terrible loss."

I noticed that Keyes's face was drawn, his eyes dull. Perhaps he acknowledged to himself the burden of guilt we shared. But at the time, my anger at him, and at myself, provided little room for sympathy.

Chastened by my experience with the Walkers, I began to inform patients at UCLA in detail about the biology of their disease and its prognosis. One of my patients was Claire Allen. A diminutive librarian in her early forties with strawberry-blond hair and two young children, she had breast cancer that had widely metastasized to her bones and liver. She had taken the news stalwartly.

We met in my clinic office. Claire sat before me expectantly.

- 4 2 -

"With this disease, Claire, a remission would ordinarily last three to six months, and a person could expect to survive between one and two years," I told her. Sharing the statistics had seemed obvious: A person like Claire surely would want to be clearly and completely informed.

She sustained her stoical disposition as she heard the numbers, and I was heartened by her show of resilience.

Claire had an excellent response to the chemotherapy. The deposits of cancer in her liver shrank until they were barely detectable on the most sensitive scans. But I later learned from her husband, as the end neared, that the months of respite from the cancer were not months she enjoyed. The statistics that I had so bluntly stated echoed endlessly in her head. She was unable to stop thinking that she could die at any moment. The points of pleasure in her day—her children, her hours reading novels— were darkly colored by the specter of death. It was as if I had cast a shroud over her, and she was forced to view all of life through it.

For a while I held on to the ideology of the "right to know." Claire Allen was not the only patient who suffered from my ineptitude. Ultimately, after venturing too far and too fast in the direction of cold facts, I was tempted to retreat behind a wall of false hope, as I did with the Walkers. I found myself giving in to this temptation, especially when caring for patients who reminded me of my own family.

Henry Gold was a short-order cook in his sixties, with advanced leukemia and a wry sense of humor like my grandfather's.

I kept telling him there were treatments to be tried, even though none was likely to help. Henry endured these toxic drugs during his long weeks in the hospital. Then his blood counts fell from the chemotherapy, and he hemorrhaged around his lungs. The interns drained the blood with chest tubes. As his breathing worsened, Henry was transferred to the ICU and placed on a respirator. I never told him the truth. He died on the respirator in the ICU, unable to speak to his family and friends.

I continued to work with Richard Keyes throughout the years of my fellowship. After Frances died, our relationship remained cordial but was distinctly cooler. Over the months I noticed a change in him that seemed to go beyond our interactions. His outgoing manner was subdued, the smile he wore for patients and staff gone from his face. The memory of Frances Walker was still fresh in my mind, but it seemed unlikely to me that the drastic difference in Richard's persona stemmed from sustained grief. Physicians, out of necessity, learn how to compartmentalize emotions, lest they interfere with the care of the living. After we saw a case that was relatively straightforward and had a very good prognosis, a middle-aged woman with a small, early cancer in her breast that could be readily treated, I noted how Richard sat for several minutes, as if drained of energy, after the patient left the examining room.

The office was unusually busy that Friday afternoon, so Richard asked me to begin seeing the next patient while he finished the paperwork on the woman with the early breast cancer.

"Come quickly!" a nurse cried from the adjoining room. "Dr. Keyes fainted."

I rushed over. He was on the floor. The nurse had put a pillow under his head and elevated his legs onto a chair. His face was pasty white, his lips pale. He was slowly regaining consciousness, and I began to examine his skull for signs of trauma.

"I must have blacked out," Richard said weakly.

"Do you remember what happened?"

"I recall my legs buckling slowly. So I braced myself on the corner of the desk before I gave way."

It was late afternoon, and Richard said he hadn't had solid food all day, and no liquids except a single cup of coffee at dawn. Perhaps that accounted for the faint.

The nurse arrived shortly with a tall glass of apple juice from the office refrigerator. But when Richard sipped it from a straw, he gagged. I quickly took the cup from his hands and turned him on his side. He began to vomit bilious fluid.

"Call an ambulance to take him to the ER," I said to the nurse. I would stay in the office and attend to the patients.

I phoned the ER and spoke with the doctor on call. Some hours later, I heard that Richard had been put on an intravenous line and given an infusion of saline while doctors drew a panel of blood tests. Richard had recounted his symptoms of the preceding weeks to the ER doctor. His appetite was poor. He'd had bouts of nausea and moments of severe abdominal pain but no vomiting. He had lost over ten pounds. It struck me that none of the symptoms were subtle, and any one of them would have prompted a person, layperson or professional, to consult a physician. Richard had not. When examined by the emergency room physician, Richard experienced pain when pressure was applied under his right ribs.

The blood tests showed an increase in both the serum bilirubin and alkaline phosphatase, suggesting that something was impairing Richard's excretion of bile. It could well be a gallstone. Or it could be some other type of obstruction, benign or malignant, within the ducts that carried the bile from the liver into the small intestine to facilitate digestion.

Richard was admitted to the hospital, and a gastroenterologist took over his care. Other physicians in the community were covering for Richard, doing the best they could in managing his cases. I telephoned Richard from Los Angeles several times early in the week, but the line in his hospital room was always busy. Finally, I called the nurses' station on the floor. I was told that Richard was resting and had asked that no calls be put through.

His office secretary called me at the end of the week and said, sobbing, that Dr. Keyes was gravely ill. "He had a CAT scan and an endoscopy," she said. "He has cholangiocarcinoma."

Cholangiocarcinoma is a relatively rare tumor that originates in the cells of the bile duct. Generally, its prognosis is very poor, since the cancer tends to grow and infiltrate the liver before symptoms appear, so that when the diagnosis is ultimately made, it is already too late for effective treatment.

"He is being transferred to L.A. for surgery," the secretary told me.

The next day Richard arrived at UCLA Medical Center. He was in a private room at the far end of the corridor. His door was closed. I knocked several times before getting a reply.

"Who is it?" he asked weakly.

I answered and entered. The blinds were drawn, the room nearly dark, illuminated only by a small lamp at the bedside. A stack of magazines was piled on a table. They looked untouched.

I drew a chair close to the bed. Richard struggled to smile.

"I'm glad you came down here for the operation," I said. "Frank Andrews is the best hepatobiliary surgeon there is."

"It hardly seems worth it," Richard said.

"What do you mean? The scans didn't show any clear evidence of invasion into the liver. There's a chance that the tumor can be fully excised."

"I like that: 'clear evidence.' You know you often don't see invasion even with the best scans, and you're still dead within a year. If it weren't for my wife, I'd call a halt to it all right now."

Richard was conceding defeat without even trying.

"Andrews is a genius in the OR. If there is a way to resect it all and cure you, he'll find it," I said.

Richard eyed me warily. "Jerry, I know the game you're playing. I've played it my entire professional life."

———

Richard's case was the first of the day on the OR schedule. About five hours after it began, I went to the family waiting room. Richard's wife, Molly, sat nervously turning the pages of a magazine. She was a short, stocky woman with gray hair cut in a pageboy and a soft, round face. The Keyeses had never had children. Mrs. Keyes worked as an elementary school teacher in Russell. Richard had told me that she had studied Spanish at Bryn Mawr and used it to good effect in a school where many of the children were from indigent Mexican families.

"Any word yet?" I asked. Richard had asked that I be with Molly when Andrews came out of the OR to report on the surgery.

She shook her head. "His spirits are pretty low," she said. Then she added, "Understandably. Day in and day out, seeing people with cancer. It takes its toll."

"It does."

"But," Molly said, struggling to rally, "maybe we'll hear good news from Dr. Andrews."

The news we heard from Frank Andrews three hours later was not good.

"It extended close to the base of liver," he began, "and while there was no sign of gross invasion, there aren't sufficient margins. We have to assume that there may be microscopic deposits we were unable to remove."

There were no proven treatments for cholangiocarcinoma beyond surgery, Andrews explained to Molly, but he had seen a few patients similar to Richard who had received postoperative radiation to the base of the liver and appeared to benefit from it.

"I admit these are just anecdotes, but it is his best chance," he said to Molly, whose eyes were welling with tears. She reached for my hand, and I held hers firmly. "I've already called the radiation therapy team to see him. Once he heals up, we can get started."

Molly thanked him for all that he had done.

———

I visited Richard the next day. His room again had little light. There was an intravenous line in his hand and a tube through his nose to suction fluid from his stomach. His cheeks were sunken due to his considerable weight loss over the preceding weeks. Before I sat down, I opened the blinds halfway. There was a commanding view east over Westwood Village. The sun cut through the haze and illuminated the tile roofs of the village shops to a glowing orange.

Richard turned his head away from the window.

"Are you comfortable?" I asked.

"I'm getting enough pain medication," Richard said.

I told him that Molly was concerned about his spirits.

"What you are going to tell me—to buck up, that everything will be hunky-dory?"

"Not really," I said. I had no specific agenda other than visiting.

Richard looked at me gravely. "It's Molly I am most worried about," he said. "The hospital will find another oncologist to take over my practice. Russell is hardly a plum place, but you can make a life there. It pains me to think of Molly alone."

"There's still a chance of cure here," I said.

"Cholangiocarcinoma, Jerry. Cholangiocarcinoma. Andrews told me what he found at surgery. The tumor extended far up in the bile duct. You know what that means."

"It means that the cancer may have spread to the liver. But it's not certain. And he has had, he says, some positive results with radiation."

Richard stared fiercely at me. "I won't be strung along," he said bitterly. "What—get dragged down to radiation every day, come back feeling sick as a dog? You know what it is to radiate the base of the liver. There isn't enough Compazine in this whole hospital to stop me from vomiting my guts out. And when that's over, everyone can say that they tried their best and then tuck me in my coffin with a clear conscience."

Richard's words hit me like a blast from a furnace. My mind went blank, and then I saw where this was leading.

Richard was going to refuse the radiation. He would tell Molly that it was pointless, nothing more than torture in the guise of modern medical therapy. He would argue that it was his right to die with dignity, meaning comfortably, without "useless" toxic treatments. That would make sense if there were no possibility of survival. But there was, albeit a small one.

"You really think that Andrews is just stringing you along? That it's all a charade?" I asked.

Richard met my questions with obdurate silence.

I replayed in my mind the months that I had moonlighted with him in Russell; the guilt I still carried about the Walkers welled up. Then my thoughts shifted to Claire Allen and other patients at UCLA whom I had treated. I had swung wildly between matter-of-factly revealing shattering statistics and hiding salient facts behind euphemistic evasions, in each case unable to locate a middle ground. I was uncertain what more to say now.

After Richard fell ill, my anger with him largely abated. But I had not forgiven myself for what happened to Frances and Sharon. From the beginning of medical school, I was told that doctors made mistakes. Our teachers assured us that it was impossible to be perfect, that during our careers we would fail to make diagnoses, would incorrectly prescribe medications, would be inept at procedures. But the belief was that by recognizing and acknowledging your mistakes, you were less apt to repeat them. One of the attending physicians at the Massachusetts General Hospital, a senior cardiologist whose clinical acumen was widely acknowledged as extraordinary, was said to keep a piece of paper in the breast pocket of his white coat with a list of all the errors he had made since his internship. The hospital also had a formal mechanism for airing missed diagnoses and discussing faulty treatments with one's colleagues. It occurred behind closed doors at what was called the morbidity and mortality, or "M&M" conference. At the conference, two kinds of errors were discussed: errors in judgment and errors in technique.

But there was no similar conference about language, no analysis of errors in judgment or technique about what we said to patients and their families. The Walkers and the Allens had shown me how powerful and penetrating a doctor's words could be. Yet I still wasn't sure how to speak and what to say.

"You remember Frances Walker . . ." I began.

"Of course."

"I spoke with Sharon when we were arranging hospice."

"Russell has wonderful hospice nurses. Molly will benefit from them."

"I'm not talking about hospice or Molly." I paused. "Richard, Sharon Walker said something to me that hasn't left me. She was bitter that we hadn't been more explicit about her mother's condition."

"Jerry, we've already gone over this territory."

"No, let me continue. She said something along the lines of we didn't think people like the Walkers were smart enough or strong enough to cope with what they faced."

"I trust you explained it wasn't racial prejudice."

"Actually, I felt so bad that I didn't explain much at all to Sharon, only that we didn't doubt their intelligence."

Richard looked away for a while, deep in thought.

What we doubted, I told myself, was whether Frances and Sharon were capable of hope—hope in the face of significant uncertainty. I had often fantasized about how I would react if I were afflicted by one of the severe diseases that I dealt with as a doctor. In quiet conversations during my internship, residency, and specialty training, I found that my colleagues shared these thoughts. Doctors like Richard and me doubted not only the resilience of our patients but also our own capacity to hope. If we did not believe in our hearts that there was something to hope for, then we either pretended there was, as we did with Frances Walker, or made explicit that there wasn't, as I did with Claire

Allen. We never gave Frances and Sharon and Claire and her husband the opportunity to choose what to hope for.

You didn't have to shatter a person by frankly citing statistics, as I had with Claire Allen. These cold numbers were handed down the way a judge designates the days to be spent on death row, leaving the person with nothing but fear. Nor did you have to sustain sins of omission, as Richard and I had done with the Walkers. The evasions, the elliptical answers, the parsed phrases were all supposed to be in the service of sustaining hope. But that hope was hollow. It was false, a seductive but only temporarily satisfying illusion. Once a physician lied to a patient, even if he told the lie with good intentions, then he could not be trusted again. A patient felt betrayed and helpless, lost in the dark terrain of her disease, without a reliable guide. As time passed and the disease advanced, the truth became irresistible, and the patient's illusions evaporated, leaving her and her family with nothing but resentment and anger, as had happened with the Walkers. Any future hope the physician might try to raise would be seen as false, even if it was true.

There was a knock on the door, and Frank Andrews entered. He was a short, lean man with reddish hair and a ruddy complexion, immaculately dressed, with a perfectly knotted tie and starched white coat.

I made to leave, but Andrews said that wasn't necessary: He was dropping in to see Richard before his next case and would return later in the evening.

Andrews drew a chair up to the bedside and placed his hand

on Richard's shoulder. "You are coming along well postoperatively," he said.

"I guess so."

Andrews paused, and I saw him study Richard's listless face. "I have spoken with Dr. Beaufort, the chief of radiation therapy. He will personally oversee your case. I work closely with him, and he is outstanding."

Richard didn't reply.

"Richard, radiation gives us a chance here," Andrews said emphatically. "We don't know whether it will work, and it may not. But it could."

I observed closely what Andrews was doing.

"So, yes," Andrews continued, "you should plan for a bad outcome—draw up a will, make your wishes known in an end-of-life directive so that if you don't want to be placed on a respirator or have cardiac resuscitation, then you won't."

It seemed, paradoxically, very easy to paint the worst-case scenario. I thought about this, how our minds naturally jump to picture the negative outcome and stall there. It is because the mind is frozen by fear, and fear overwhelms hope.

"But there is a best-case scenario, Richard, and it's a real one. Speaking completely honestly, the odds are against it. Still, the radiation could destroy the cancer cells that are probably there. The treatment *could* cure you."

Richard shook his head doubtfully. "I don't know, Frank. I just don't know. The idea of cure . . ."

"It may not happen. But it *could* happen. Or the radiation may slow the growth of the cancer, not completely eradicate it, but slow it long enough so that it buys you time—time to share with Molly, time to continue working. Or, if we're lucky, it may buy enough time until a better treatment is developed."

Andrews was being truthful. Oncology was beginning to change. It wouldn't change overnight, but even in the few years since I had finished medical school and entered my fellowship, the tide had turned for previously devastating diseases, such as testicular cancer. Widespread testicular cancer had been as fatal as any tumor, including cholangiocarcinoma. It routinely killed men within a year. In the late 1970s, when I finished my residency, a new drug called cisplatin was developed. It was a derivative of the metal platinum. The effects of cisplatin in men with testicular cancer were extraordinary. In patient after patient, this new therapy melted away large, bulky masses of the tumor that had spread to their lungs and liver and bones. Within months, these men who had stood on the brink of death were returned to a full and active life. I felt as if I had witnessed a miracle, and indeed I had—the miracle of medical science turning the miraculous into the commonplace. Men with testicular cancer were no longer consigned to death. Life, rather than death, had become the rule.

Andrews glanced at the clock on the wall. "We need to speak more, and in greater depth. But this is not an open-and-shut situation. I wouldn't recommend that you go ahead with radiation if I didn't think it could really make a difference. It has in other cases of mine, and I'd like to count you in that group."

Andrews left. I saw how he had purposefully left Richard with hope, slim but true, and I imagined that when he returned, he would continue to reinforce the possibility of a cure.

I stood to leave. Richard thanked me for visiting and then fixed on my eyes.

"Do you believe him?" he asked.

"Yes, I do."

Richard had a difficult time with the radiation treatment. It was months before he had the energy to work even a few hours. He had anticipated this, and while I could moonlight on weekends, it was clear that Russell needed a full-time oncologist. Richard recruited a young woman from San Diego as an associate. She was willing to take the risk and shoulder the burdens of living and working in Russell. Richard told her there were hidden joys that would soon become apparent.

The cloud of cancer hung over Richard and Molly for many years. As a rule of thumb, if a tumor like cholangiocarcinoma does not regrow within five years, it can be said to be cured. Richard went regularly for checkups and scans with Dr. Andrews. I left UCLA in 1983, shortly after the birth of my first child, Steven, feeling the powerful pull of family in Boston. On the sixth anniversary of Richard's surgery, I received a telephone call from him.

We chatted about Molly, how she hadn't slowed down despite

nearing retirement. Richard worked three days a week, spending more time at the piano. I told him that Steven was sprouting, and Pam and I had just had our second son, Michael.

As the conversation drew to a close, I said it must be an indescribable relief for him to finally put the cholangiocarcinoma behind him.

"It is. But the experience of being a patient is not something I'll ever put behind me." It had profoundly changed him, Richard said, for the better.

I did not probe further, but I sensed that in his clinical practice, and in his own life, Richard had come closer to the middle ground where both truth and hope could reside. I had seen that it was possible.

CHAPTER 3

The Right to Hope

O ne evening in the summer of 1987, when I was a good
way further along in my career, I went to visit another ill
colleague, just as I had visited Richard Keyes. George Griffin—
Harvard professor, revered and beloved chairman of the Department of Pathology—had been diagnosed with stomach cancer.
The bitter irony was that stomach cancer was the disease that he
had made his life's work. No one knew more about the malignancy and its prognosis than George. As I exited the elevator on
the tenth floor and headed to the oncology unit, Eric Paulsen, a
fellow cancer specialist, intercepted me and led me to the far end
of the corridor, out of earshot of the nurses and other staff.

"It's madness, pure madness, what George is doing," Eric said.

George had insisted on receiving aggressive treatment combining high doses of chemotherapy with intensive radiation, de-

spite the absence of any evidence that such toxic therapy could change the fatal outcome of a cancer as advanced as his. Neither Eric nor I would ordinarily treat a patient with George's prognosis this way. But we were not consulting on his case. Rather, a senior oncologist of George's generation had been called in.

I had no explanation; I told Eric that I, too, couldn't fathom why George had chosen this path and had seemingly convinced the oncologist to pursue it with him.

After Eric left, I stood deep in thought. George's choice to be aggressively treated not only seemed like a desperate, wrong-headed, ultimately useless effort to resist the inevitable; it also appeared to be a blanket contradiction of who he was and how he approached life and death.

More than any other medical specialist, a pathologist grasps the inescapable reality of death. It is the pathologist who conducts the autopsy, removing, one by one, the liver, kidneys, heart, brain. It is the pathologist who sits before the microscope viewing the inner universe of blood vessels and muscles and nerves and glands that make each organ what it is. No doctor is more intimate with the substance of our selves. And each organ and tissue the pathologist studies serves as palpable testimony that we are material and mortal.

I wondered whether George's intimate knowledge of death had triggered an extreme form of denial—so extreme that he was subjecting himself to iatrogenic torture. He risked hastening his demise, or at least robbing himself of the last tranquil days at home with his wife, his children, his friends. What Eric Paulsen termed "madness" seemed rather a sad, self-defeating loss of judgment.

I entered George's room. The sheets were drawn to his neck. His eyes were closed and sunken, his skin ashen, his lips blackened from dried blood in deep ulcers. I wondered briefly if George had died and I was the first to find him. Then he slowly turned his head and noticed me. With tears in his eyes, he struggled to speak, but all he could muster was a harsh, unintelligible whisper. "Don't talk," I said. His eyes closed in assent.

George had suffered a severe side effect of the treatment, essentially a burn of his alimentary canal. The delicate tissue from his lips to his rectum was scorched, ulcerated, and bleeding. The only ameliorative measures were morphine and intravenous fluids.

I spent a few short minutes in the room. I worried that George would feel compelled by his good nature to try to engage me in conversation and further exhaust himself. Typically, as I departed from such a visit, I would say to a patient how vital it was to keep fighting against the cancer and emerge whole. I offered such encouragement to people with leukemia, lymphoma, testicular cancer, and other malignancies that, despite their severity, could be cured. Sometimes the pain and suffering from these diseases and their therapy felt like a descent into hell. As a physician, I knew that in these cases there was a possible exit, that if the patient could endure the harsh treatment, then the leukemia or lymphoma or testicular cancer might be eradicated and life could be restored. I had learned through my years in practice that it was the role of the doctor to always keep before a patient's eyes the fact of this exit, that reaching it was feasible, not guaranteed but certainly a realistic possibility. With steady hands and strong words, the doctor supported his patient's body and spirit through the vicissitudes of the therapy to that door to a cure. But in George's case, such words seemed hollow, empty of any truth. They would speak only false hope, denying the reality of metastatic stomach cancer, a reality that George himself had defined. Instead, I left with platitudes, saying how much everyone missed him, how we all were thinking of him, and that soon he might find some relief.

For days I could not get George out of my mind, nor stop speculating why he had subjected himself to such extreme treatment. I still wanted to ascribe his choice to denial, common among patients in severe circumstances, but that didn't seem to be the answer based on George's recent behavior. Days after learning his diagnosis, George had relinquished his clinical duties and explained to the CEO of the hospital why he was stepping down. He told the head administrator exactly what his chances were, what the future held: His type of cancer, he said, made it unlikely that he would live longer than six to nine months.

George's choice of treatment was further perplexing because he was not only a scientist but a person of deep faith. He was descended from New England Puritans, and his family had migrated west in the early 1800s, among the early pioneers who reached the Pacific Coast. The Griffins had held fast to their Congregationalist creed, and George was raised to view this world as a prelude to the next. The body that lay on the autopsy table was but a temporary vessel; the soul was immortal and at death sought to be reunited with God. George would readily acknowledge that all of us have a natural fear of death, but his belief in God and in an afterlife assuaged it.

Could the senior oncologist overseeing George's case have persuaded him to undertake this aggressive course of treatment? It seemed unlikely. George knew the clinical terrain better than anyone else, and no one could mislead him with false data or inflated promises. No, George had plunged himself into this abyss of suffering, which would reach its end only in death.

I sought to replace the awful images from that afternoon's visit with memories of George Griffin in full health. Sixty-one years old, a wiry man of middle height with thinning gray hair and sharp features, he was conservative in his dress and understated in his demeanor. His great delight was the travel that his research projects afforded him. He had ventured into the most remote corners of Asia to study the epidemiology of stomach cancer, seeking the genetic and environmental factors that caused some populations to suffer from it frequently and others to be afflicted rarely. He had forged friendships with fellow researchers and become expert in their cultures. George was particularly interested in Korea and rural China. This interest deepened after his first wife died of colon cancer and George married Eunha, a Korean pathologist.

George was much esteemed in the hospital. I regularly attended the pathology conferences where he presided. He would listen with laser focus to each physician's opinion about complex and controversial cases, and after all had aired their thoughts, would concisely analyze the issues, weaving together a tight argument for one particular diagnosis. His conclusions seemed to leave no loose ends; his judgments rested squarely on facts.

Staff physicians like me particularly looked forward to conferences where cases of stomach cancer would be discussed. George could expound like no one else on the Harvard faculty about subtle nuances of the disease, pointing out characteristics in the architecture of the patient's particular tumor that revealed whether its behavior would be indolent or aggressive. George peppered his dissertations with colorful stories from his travels, tales of remote villages in the mountains of northwest China, or teeming cities in the basins of South Korea; woven in was the medical history of gastric cancer. George emphasized how his work never could have been accomplished without predecessors over the last century who had found puzzle pieces to explain why some people developed the tumor while others did not. Nor did George fail to credit peers who published important articles that went one step beyond his own work. Not a single fact or hypothesis about stomach cancer seemed unfamiliar to George, and his encyclopedic knowledge, provided in his characteristically understated way, never failed to impress us.

Just a few months before, in the spring of 1987, George had noted a foul taste in his mouth. Even small amounts of food and liquid made him feel bloated. George had never experienced digestive problems before, and he had not lost weight or noticed a decrease in his appetite. He sought the help of Dr. James Pierson, a senior internist at the hospital. A physical examination showed no abnormalities. Still, antacids only temporarily relieved the unpleasant taste. Then, one day at lunch, after a few bites of food, George vomited. Dr. Pierson ordered a barium swallow exam. Barium is an opaque, chalky liquid that, when swallowed, outlines the interior of the esophagus and stomach on X ray. At the entry to George's stomach was a mass the size of a clenched fist. A CAT scan showed a cluster of enlarged lymph nodes encircling the mass, like moons around a planet.

The mass and surrounding lymph nodes almost certainly represented cancer, but which kind was unclear. Some tumors of the stomach are easily treated and can be cured. Others, according to George's own research and the studies of others, are resistant to all therapies and are routinely fatal. A biopsy of the mass was taken to make the diagnosis.

George had asked his wife to view the biopsy tissue with him. Eunha, upon peering into the microscope, burst into tears. George didn't flinch. There were sheets of large distorted cells, many in the midst of dividing. Some cells had grown beyond the lining of the stomach and were invading the organ's blood vessels and muscular wall. George paused, then said in his even voice that it was a "classic case" of undifferentiated carcinoma. He did not need to add that it was the worst type of stomach cancer that one can have.

Stomach cancer is the number two cause of cancer deaths worldwide, and it strikes some twenty thousand Americans every year. Men develop the disease more often than women, in a ratio of about two to one. Environmental factors, particularly certain foods and dietary practices, appear to contribute to the genesis of the malignancy. These include preserving meats and fish by smoking and salting, often without refrigeration. This is seen most clearly in studies of migrating populations that change their eating habits. For example, Japanese families who move to Hawaii and maintain their traditional diet have no decline in the high incidence of stomach cancer, while those relocated Japanese families who change to a Western diet have a markedly reduced risk. Similarly, Polish immigrants who live in the United States for ten years or more and generally adapt to American customs show a significant decrease in their risk for stomach cancer as compared with Poles from the same locales who stayed in Poland. In both the Japanese and Polish groups, eating fewer preserved smoked or salted foods as part of a modern American diet appears to mitigate their risk for the malignancy.

Genetics also plays a role, and certain families appear to be particularly susceptible to the dietary triggers of stomach cancer. The most notable genetically affected family was Napoleon Bonaparte's: The emperor, his father, and his grandfather all died from stomach cancer. This interplay of diet and genetics suggests that some people are less able to break down and detoxify ingested carcinogens that damage the cells of the stomach and transform them into a malignant state. Nitrates appear to be the primary carcinogenic culprits in preserved smoked or salted meats and fish. Laboratory animals fed diets rich in nitrates develop cancerous lesions in the gut.

Although George Griffin had traveled frequently in Asia, he had never favored the diet of the lands he visited. Eunha had arrived in the United States for her medical training several years before she met George, and she rarely prepared Korean cuisine containing heavily smoked or salted foods. No one else in George's family had had stomach cancer.

While it remained a mystery why George had developed stomach cancer, it was brutally apparent what he faced. In cases like his, as he told the hospital CEO, only 2 to 3 percent of patients live six months. At nine months, survival is less than 1 percent.

Shortly after the diagnosis, and before he started on the treatment that resulted in his hospitalization, I visited George in his office in the pathology department. He greeted me warmly. I had come for help in diagnosing the complex case of a patient transferred from another hospital. George mounted my patient's biopsy slide on the stage of his microscope and directed me to a connected pair of eyepieces so we could look at it together. He moved the lens over the tissue, pointing out the salient features that would form a diagnosis. "A bronchoalveolar carcinoma," George said, naming a lung cancer that often arises in scar tissue and can masquerade as pneumonia, filling the airspaces with malignant cells and inflammatory fluid.

George moved away from the microscope. He blinked a few

times, then spoke haltingly. "It can't be resected at this stage. As you know, the prognosis of bronchoalveolar carcinoma is very poor." I heard the strain in his voice. "You also know the prognosis I face. The administration of the pathology department will be assumed by Dr. Tibor. He's a good man. He'll be here when you need him." I nodded and acknowledged what did not need to be said.

It took three weeks for George's chemical burn from the high doses of chemotherapy and radiation to subside. His appetite was still poor and his tolerance of food limited. He ate small portions and felt queasy with liquids that were too cold or too hot. He had lost nearly thirty pounds; the flesh of his face was pulled taut over his cheekbones, and the veins in his neck protruded. Resisting Eunha's pleas to rest at home, he went back to work, sitting in his office two or three hours a day, shifting papers and studying an occasional slide. He moved about his department like a specter.

Not long after the therapy's toxic effects had waned, I learned that George was back in the hospital, scheduled to undergo surgery. The news moved like a shock of electricity through the clinical staff. "Cowboys, goddamn cowboys," my oncologist colleague Eric Paulsen said bitterly of the impending operation. In medical vernacular, a "cowboy" is a trigger-happy practitioner, always at the ready to cut or prescribe, even when there is no rationale. "Why not just put a gun to his head?" Paulsen exclaimed.

Surgery was clearly indicated in a case like George's when the cancer was obstructing the esophagus or stomach, or there was uncontrolled internal bleeding. In these instances, the blockage needed to be removed or the bleeding vessel in the tumor tied off. But George had neither of these complications.

A CAT scan done after his radiation and chemotherapy showed there had been some decrease in the size of the mass and surrounding lymph nodes. This was not unexpected. Toxic treatments often result in some shrinkage of the tumor, a so-called partial response. But such changes in response to radiation and chemotherapy are transient and do not alter the ultimate outcome. Moreover, a surgeon's scalpel could not remove the multitude of cancer cells that already had entered George's blood vessels and penetrated beyond the lymph channels, and were deposited throughout the abdomen like buckshot. These microscopic clusters of tumor, which could not be identified with even the most advanced techniques, would be left behind after surgery. Within weeks they would regrow and spread again. So the trauma of major surgery—general anesthesia, blood loss, complications of infection and blood clots—only risked stealing more days away from the few that remained for George.

George's was the first case on the OR schedule. Eunha had spent the night at his bedside in prayer. The surgeon made the initial incision at the lower end of George's sternum, the breast bone, and cut down to the crest of his pelvis. He needed an extensive field to operate in, given his expectation that the cancer had spread widely. The stomach is shaped like a Portuguese wine pouch: Its tapered neck receives the chewed food from the long tube that is the esophagus, then churns the meal by power-ful muscular contractions in the large curved middle, or the fun-dus. After the food is broken down into a thick slurry, it moves to the lower end of the stomach. At the exit is a muscular ring that opens and closes like a spigot to regulate the steady passage of the digested contents into the small intestine.

The gloved hand of the surgeon moved inch by inch along George's exposed stomach, palpating the outer surfaces to define the limits of the mass. It was a large tumor—that much was known before the operation from the X rays and the endoscopy. To resect it, he identified each of the arteries that supply blood to the stomach and clamped them shut. Then he traced the web of interlacing nerves that cause the stomach to contract and relax. These were cut. The organ was now isolated. The surgeon moved his scalpel along the thick chain of lymph nodes that sur-rounded the stomach, excising each cluster of nodes and deposit-ing them onto a bed of sterile gauze. A pathologist stood at the surgeon's side to receive the tissue. Ivory white, ranging in size from a small pea to a walnut, the lymph nodes must have filled the pathologist with dread. They were irrefutable evidence of lethal metastasis, the march of George's cancer into his abdomen and beyond, making him, by standard criteria, incurable.

The surgeon then sliced along a wide arc beginning at the lower stomach, along the rounded fundus, and finally reaching the organ's neck. The cancer had extended its tentacles into the esophagus, and its lower third had to be sacrificed. George would not be able to eat without a conduit from the upper esophagus to the residual stump of stomach. So the surgeon swung up a piece of intestine from the abdomen and grafted it between the remaining esophagus and stomach stump to serve as a surrogate channel.

Seven hours passed before the surgery was over. George lost liters of fluids during the operation and left the OR with blood and plasma and saline IVs attached to him. He was sent directly to the ICU for monitoring. His colleagues in the pathology laboratory began their work with his removed tissue.

Dr. Tibor, the acting head of the department, directed the examination. The stomach was splayed open. The cancerous mass was coal black in color, its surface marked by large craters. Dr. Tibor gently moved a metal probe over the edge of a crater, and flakes of blackened tissue broke off. The excised lymph nodes were similarly pocked black. "Extensive necrosis," Dr. Tibor noted on his report—a technical term indicating widespread death of tissue, which accounted for the coal-black color. Necrosis was expected after the treatment, given the burn George's gut had suffered from the chemotherapy and radiation. Dr. Tibor cut into the center of the tumor. Thin, watery blood and lymph seeped out of the furrow. The pathologist cut slice after slice of the mass, searching for nests of living cancer, but no viable tissue was evident. The sections of necrotic tissue were fixed and stained and then studied under the microscope. Dr. Tibor expected that tracts of tumor would be found under the magnification of the microscope. But the cancer cells were nowhere to be seen.

The textbook explanation for the failure to find the cancer was that it was hiding. Many cells had been destroyed by the radiation and chemotherapy, but many would be impervious, even to this assault. And those cancer cells were not static. They circulated in the blood and lymph, traveling through the body until they found new residences. Some tumors are notorious in this regard, like breast cancer or melanoma, in which the primary lesions appear to be eradicated only to pop up weeks or months or even years later with metastatic deposits in the liver or bones or lungs or brain. With such an aggressive type of stomach cancer as George's, new masses would soon appear, likely in his abdomen, then in his lungs. The cancer ultimately would fill his intestines and occupy his chest until it killed him.

But it seemed that George would not acknowledge this outcome, even if it was based on a biology he had spent his life studying. Not long after the surgery, he was admitted to the hospital to receive yet more chemotherapy and radiation. Along with Eric Paulsen and most of the senior staff, I was deeply saddened by George's decision to undergo even more toxic treatment that seemed pointless. There was some change in the treatment regimen: The dose of radiation was decreased and George was spared the drug that had been the primary culprit causing his chemical burn.

Not long after this second course of chemotherapy, I visited George. It was again at twilight, when the cacophony of the hospital day—the chatter of doctors and nurses, the steady steps of visitors in the corridors, the groaning gurneys moving through

THE ANATOMY OF HOPE

the halls—had given way to the steady hum of fluorescent lights and the cricket clicks of heart monitors and infusion pumps. I knocked softly on George's door and entered. In the dusk, his skin had an almost metallic cast. His cheeks were more sunken than when I had last seen him a few weeks before. He offered his thin hand. It was warm, and I suspected he had a fever. Blisters again pocked his lips. An infusion bottle hung on a pole attached to the bed. I glanced at the label on the bottle and saw that it contained morphine.

Eunha sat at his bedside with a Bible open on her lap. She closed it. "It's kind of you to come," she said with a soft smile. She was dressed in a navy blue suit, and her rich black hair was perfectly coiffed. But her eyes were etched with sadness, and her hands tightly gripped the closed Bible.

I embraced Eunha. She told me that the anticipated side effects of the treatment had occurred: George's blood counts were low, and his gastrointestinal tract was again inflamed and ulcerated from the chemotherapy. The pain had been relentless until he at last broke down and accepted morphine.

I tried to speak to Eunha with my eyes. As a physician, she knew the same painful truth that we all did. Why continue? She looked away.

If I were George's doctor, I told myself, I would have taken Eunha aside and delicately but firmly guided the conversation to question his decision to resume chemotherapy and radiation. Wasn't that the doctor's role, to protect George, and to some degree his wife, from what seemed to be a stubborn descent into more pointless treatment? There had to be an end to this torture, short of his death. But I was not his doctor. It was not my role to intervene. So I masked my frustration and pity and told George and Eunha that they were in my prayers.

On a chilly December day in 2000, thirteen years after George's diagnosis of stomach cancer, I sat in the atrium café of the hospital. I had chosen a table under the skylight, and the winter sun cast an indistinct rainbow on the steel table. I rose from my chair as I spied George. He greeted me warmly.

He and Eunha had retired to southern New Hampshire after his last treatment. It took nearly a year for George to tolerate solid foods without pain. When his strength returned, he began to travel to Boston one or two days a week to consult on controversial cases. But as time passed and the cancer did not reappear, it seemed that he was cured. George decided then to devote himself to church and community activities, and to traveling with Eunha.

I had never summoned the courage to speak with George about his decisions. He was a private person, and I wasn't certain how this rendezvous to discuss his case would unfold. My reluctance, I acknowledged, was more than concern about the borders of privacy. I carried a gnawing sense of guilt. I and the rest of the clinical staff had written George off. He would not be alive if our recommendations had been heeded.

George chose his meal judiciously: a bowl of chicken noodle soup, some slices of white bread and butter, and a tall glass of apple juice. Large amounts of bulk and fiber were still difficult for him to digest comfortably, he explained. The interposed intestine and remaining stump of stomach were poor substitutes for normal organs. He was still gaunt, and the collar of his button-down oxford hardly met his neck. Even after thirteen years, he had not regained all of the lost weight. Still, his eyes were animated and his voice was strong, and he ate the little he could with relish. We exchanged news about our families and bits of hospital gossip until it seemed time to address the reason for our meeting.

I began at the end, in the autumn of 1987, following the radical surgery and the second course of chemotherapy and radiation. Had he been able to insist on these treatments by ignoring the grim statistics and simply telling himself he would survive?

"Even when I was down at the bottom," George said, "I knew exactly what the numbers were." He had never had a moment of denial or delusion. When he returned home from the hospital, during the last days of fall, he assumed he would soon die. He and Eunha were avid gardeners, and he asked her to go to the nursery and purchase daffodil bulbs. Weak as he was, George planted them himself. The daffodils would bloom the coming spring. "I told myself: Maybe I'll get to see them flower, but likely I won't. Then they will be for my grave."

George's face was calm and expectant. The message was that I should speak with equal candor. "Did you know that I and virtually the entire medical staff disagreed with the treatment?" I asked.

"I did." George paused gravely. "I knew all the arguments made in cases like mine. Treatment would cause unnecessary suffering—for me and for my family. Add in that it throws away society's money on a doomed person." His eyes narrowed. "I find these arguments patronizing. I did even before I was ill. Most patients don't really understand what's happening to them, how poor their prognosis is, because they're not clearly told the odds by their doctors. Rather, the usual course of their disease is described in oblique or misleading terms. I, of course, had a crystal-clear understanding of my chances. And it was my *right* to choose what I did.

"Even if I didn't prevail—and I didn't expect to—it was my only chance. I deeply wanted to live, so I had to fight. Then I could tell myself that I had tried, that I had done everything possible. There would be no regrets."

His was a libertarian mind-set, one that placed the individual squarely as the ultimate arbiter of his fate. It represented a certain form of hope—the hope to be strong enough not to yield, to have the determination and the fortitude to fight, despite knowing that there was little chance of survival. The mustering of the will to engage the foe and the strength to sustain the battle, in themselves, became a form of victory. Surrender would be on George's terms, at a time and place of his choosing. I realized how clueless I had been.

"Once I decided to go for it," George continued, "I thought of my forebears. They were pioneers who embarked on a journey west that was perilous and uncertain. Most knew they would perish on the way. But they persisted."

And his faith, I asked, how did that influence him?

"I recited the Twenty-third Psalm—before, during, and after each treatment. It spoke so beautifully, so directly, to my plight."

As I recollected the psalm, it became all the more clear to me how George had sustained his fight. There are few words that so capture the courage and comfort that faith provides.

> *The Lord is my shepherd;*
> *I shall not want.*
> *He maketh me to lie down in green pastures:*
> *he leadeth me beside the still waters.*
> *He restoreth my soul:*
> *he leadeth me in the paths of righteousness*
> *for his name's sake.*
> *Yea, though I walk through the valley of the shadow*
> *of death,*
> *I will fear no evil: for thou art with me;*
> *thy rod and thy staff they comfort me.*
> *Thou preparest a table before me in the presence*
> *of mine enemies:*
> *thou anointest my head with oil;*
> *my cup runneth over.*
> *Surely goodness and mercy shall follow me*
> *all the days of my life:*
> *and I will dwell in the house of the Lord for ever.*

George also said he had derived great comfort from knowing how many people were praying for him. The distances he had traveled in his career, and the diversity of the department he led, had brought him into contact with physicians and scientists of many faiths. Christians, Jews, Hindus, Buddhists, Muslims all had him in their prayers.

I, too, had prayed for him, but I had not dared to pray for a cure. It seemed presumptuous, too much to ask of God. Rather, I had prayed that George not suffer. And I told him this.

"I am a scientist," he replied. "It is hard to be both a person of faith and a person of science."

I understood what George meant. Modern science made no room for unprovable beliefs. And at the time, the medical establishment did not consider faith a significant factor in the outcome of health or disease.

But, George explained, Eunha had looked to God for more. "She had this sixth sense that I would be cured. I didn't. My survival has made her more devout, almost mystical. She was always convinced that I'd live. But scientifically, that was so remote. Like you, I didn't see the end that way."

Then how could we ultimately understand his case?

"It is a medical miracle," George said, emphasizing the word "medical." "What it says is that even the most aggressive and gruesome cancers can sometimes be stayed. They rarely are, and what we have to throw at them is so crude and so toxic. But sometimes that primitive therapy works. The unexpected happens. Of course, as a scientist, I want to explain the unexpected,

I want to know *why*. If we could answer that question, then one day we could create new therapies with the knowledge and make what seems miraculous commonplace."

George told me that for many years after his cure, he visited cancer patients in the hospital who were losing hope. He was an inspiration. His survival showed them that there is inherent uncertainty in the behavior of even the worst diseases. But the hope that George wanted to spark went beyond this clinical truth. He sought to assist people in making choices that addressed their own particular needs, desires, and beliefs. As he had learned in his struggle, such choices often were made by reaching back to the touchstones of one's heritage and faith. In George's case, it was the meld of the puritan with the pioneer. It was a search for freedom, both a freedom of the body shackled by disease and a freedom of the spirit to assert its dignity, or having some level of control over one's life.

The psalm that George cited is a touchstone to many of us. Only recently did I appreciate that its power comes in part from the subtle way it depicts God's translocation during times of need. This liturgical insight came from the rabbi of my synagogue, William Hamilton. Rabbi Hamilton pointed out to me that the psalm suggests two very different scenes. The first is an idyllic one: green pastures and still waters. The second is a harrowing one: the valley of the shadow of death. In the idyllic setting, God is somewhat distant and is spoken of in the third person: "he" maketh me to lie down in green pastures; "he" leadeth me beside the still waters. But then there is an abrupt shift in scene and of God's person. When the Psalmist walks through the valley of the shadow of death, God is addressed as "thou." The change to the familiar indicates that God has moved near to us, become a close companion, allaying our fear, supporting us with rod and staff, nourishing us in the presence of our enemies. Faith is most powerful when God is felt as proximal and personal.

———

Oliver Wendell Holmes, a nineteenth-century Boston physician, poet, and essayist, cautioned: "Beware how you take away hope from another human being." A physician should never sit like a judge over a desperate patient and hand down a fixed sentence of days or weeks or months of remaining life, even when the patient expects it.

Holmes's admonition is rarely heeded. It took George Griffin to teach me that omniscience about life and death is not within a physician's purview. A doctor should never write off a person a priori. At the moment of initial diagnosis, closing off options and denying choices is premature and clinically wrong.

This has been brought into sharp focus for me over the past decades by other "miracle" cures I have witnessed. Eva Schumer was a feisty matron with an earthy sense of humor and a taste for high-stakes poker. "I am a tough broad," she told me when we met, her raspy voice signaling decades of smoking. She had lung cancer that had spread from its initial site through the lacy fabric of the organ. A surgeon removed the tumor and surrounding tissue as well as he could, but it defied complete excision. I explained to Eva that chemotherapy and radiation might retard the growth of the cancer for some time but were not curative. "This is the worst hand I've ever been dealt," she flatly replied. After a moment's reflection, she said she could not walk away from the table. The treatment was a "bluff," she opined, that one day would be called.

While I welcomed her bravado as a sign of her will and resiliency, I explained that we might do more harm than good, that the treatment held complications that could impair the quality of her life or even shorten it. Eva insisted we go ahead. She had nausea, vomiting, hair loss, and fatigue that lasted two weeks of each month for over half a year. But her tumor steadily shrank until it could not be detected by the most sensitive scans. It never reappeared. Eva died not long ago, at an advanced age, nearly twenty years after I met her. Her hair was still dyed a flaming red, and her children were still driven to distraction by her reckless gambling.

Catherine D'Angelo is another woman whose disease did not seem to have read the textbook. In 1986 she developed a type of breast cancer called "inflammatory," meaning that the cancer cells have widely infiltrated through the skin and into the chest wall, irritating the tissue and turning the flesh crimson. Survival with inflammatory breast cancer is at best one to two years. Catherine was treated like Eva, palliatively, with local radiation to contain the cancer gripping her breast. I also prescribed tamoxifen, since the tumor was sensitive to estrogen. Catherine shared my limited expectations. She understood that the treatment might relieve the painful burning over her chest and restore some quality of life but was unlikely to gain her significant time. The inflammation subsided a few months after the therapy. For seven years, the tumor, though present, did not grow. Then a lump appeared in her armpit. It was the breast cancer. We switched from tamoxifen to another hormonal agent, Megace. The mass retreated for several years, then regrew. I prescribed yet another hormone blocker, Arimidex. This afforded another remission.

"We're playing hide-and-seek with it," Catherine said. It never again infiltrated and inflamed the skin; rather, it seemed to have changed its biology, growing as small pea-sized tumors around her breast. Each time, thankfully, by finding the tumor just as it recurred, we were able to banish it with well-tolerated drugs. Strictly speaking, Catherine is not cured. There are tumor cells lurking in her body. But her clinical course contradicts the trajectory that inflammatory breast cancer is supposed to follow. Like George Griffin and Eva Schumer, Catherine did not surrender from the start, and not surrendering has sustained her life.

I once hesitated to recount such anecdotes to patients and their families for fear of raising false hopes. Those who understand their condition and then choose George's path are not mad, as Eric Paulsen said, nor in denial, as I once believed. I have come to appreciate that the decision is not as simple as that. My place is to provide choice and understanding. To hope under the most extreme circumstances is an act of defiance that, as George explained, permits a person to live his life on his own terms. It is part of the human spirit to endure and give a miracle a chance to happen.

CHAPTER 4

Step by Step

During the 1980s, most of my patients died from their diseases. As a hematologist-oncologist, in addition to caring for people with leukemia and lymphoma and breast cancer, I served in the AIDS clinic. HIV infection rapidly destroyed the immune system, rendering the afflicted vulnerable to life-threatening infections as well as malignancies like Kaposi's sarcoma. There was little we could do to inhibit HIV, or to stave off fatal complications for long. The typical survival from the onset of full-blown AIDS was a year or two.

My respite from this misery was my laboratory. The science required intense focus and intellectual engagement. There, in calm and quiet, genes were decoded, proteins analyzed, and cells studied in their intricate choreography of growth and maturation. My spirits were briefly bolstered by the sense that one day what we learned might alleviate the suffering that seemed unchecked. But after hours in the lab, I would don my white coat and return to the grim reality of the wards. It was hard to sustain hope in the face of so much death.

Then, in the early 1990s, my clinical practice began to significantly change. Many more of my patients were living than dying. George's words about new knowledge in medicine turning the miraculous into the commonplace rang true. With each scientific advance, hope became more accessible. The maladies whose very names—leukemia, lymphoma, and AIDS—bespoke a rapid and often terrible death for the majority of the afflicted were yielding in the face of new treatments. Acute promyelocytic leukemia, a blood cancer that causes patients to lose the ability to clot so they succumb to massive hemorrhage, was found to be highly responsive to novel drugs called retinoids. Retinoids, which are derivatives of vitamin A, differed from classical chemotherapy agents in that they potently and selectively killed the leukemic blood cells and largely spared the normal ones. The prognosis for these patients went from grim to good. The outcome of AIDS also dramatically turned around with the development of combination treatments, antiviral "cocktails" that shut off the virus and allowed the immune defenses to recover. While before, I lost ten or more of my patients each year to AIDS, the new drugs made death a rare occurrence. Similarly, the downhill course of certain types of lymphoma was radically reversed. Treatment regimens melding novel chemotherapy drugs with radiation and monoclonal antibodies—those designed in the laboratory to specifically target the lymphoma cells—were able to completely eradicate some cancers and achieve a permanent cure.

I had learned from George that every patient has the right to hope, despite long odds, and it was my role to help nurture that hope. As medical science advanced, it became easier for me to help my patients see and sustain hope because I believed in it myself. A physician-scientist who witnesses discoveries at the laboratory bench that are translated into treatments at the patient's bedside can share this vision with a sufferer. But then I met a patient named Dan Conrad and found that it took more than science to make hope real.

Dan was a construction worker on a project in downtown Boston, and one late afternoon shortly before Christmas 1995, he became short of breath while hauling a load of wet cement. He had not been feeling good for several weeks, his appetite poor and his energy low. Dan sat down, expecting the breathlessness to pass, but it didn't. The foreman insisted he go to the emergency room. The ER resident examined him and heard harsh wheezes throughout his lungs. Dan had a low-grade fever, and the doctor wondered whether it might be the beginning of pneumonia. The chest X ray showed otherwise. Accounting for his symptoms was a mass in the center of Dan's chest, the size of a grapefruit, pressing on his airways.

Dan was admitted to the hospital. A CAT scan showed that the mass was not confined to his chest; it snaked like an alien invader down through the diaphragm and filled the abdomen, nearly thirty centimeters (about a foot) in diameter, encasing major blood vessels, the aorta and vena cava, with its tentacles. The next day, a biopsy was performed. It showed an aggressive type of non-Hodgkin's lymphoma. Lymphoma of this kind grows quickly, and the tumor was threatening to choke off Dan's bronchi or strangle his blood vessels or bowel. He needed to begin treatment immediately. I was called and asked to assist in Dan's care.

It was just after eight A.M. The ward was already in full swing. Nurses charted vital signs and dispensed medications. Clerks on the telephone scheduled procedures. A team of residents stood in the corridor around the chart rack, making their work rounds. I greeted the residents and told them that I would be supervising Dan's case.

"It's a huge mass," the senior resident, Virginia Chu, said. She was in her late twenties, with shoulder-length hair and steel-rimmed glasses. "Have you seen the CAT scan?"

I said I had. "Anything I should know before I go in and meet him?" I asked.

Virginia took Dan's chart from the rack and briskly outlined his status. His fever was hovering around 100°F. He'd had a poor night, unable to lie flat in bed because of the shortness of breath, so he sat up with pillows propped behind him. He had been put on an oxygen mask, but it proved of limited help.

"It's positional," Virginia said. "Once he's flat, the mass presses down, and even with the mask, he can't get enough oxygen."

"Any sign of tumor lysis?" I asked. Tumor lysis refers to the rapid breakdown, or "lysis," of a bulky, fast-growing cancer. It can happen when the tumor outstrips its blood supply and the cells, starved of nutrients and oxygen, quickly die en masse. Tumor lysis also occurs as a result of treatment, when the chemotherapy or radiation triggers widespread breakdown of the cancer. In either case, the numerous cancer cells dying at once flood the bloodstream with their toxic contents. Acids released from the dead cells disrupt the body's delicate metabolic balance, causing seizures or shock, and released ions like potassium can cause the heart to stop beating.

"No, we've kept him pretty well in balance," Virginia said. "So far."

I told Virginia and her team that I would outline the treatment with them after examining and talking with Dan.

I had read Dan's ER record online in my office before coming to the ward. He was forty-three years old, born in a small town outside Birmingham, Alabama. There was no family history of cancer and no specific risk factors for lymphoma, such as work with toxic chemicals or radiation, or infection with HIV. At the age of nineteen, he had enlisted in the army and was sent to Vietnam. During his tour of duty, Dan hadn't been in known contact with any defoliants or other substances that might have been carcinogenic. After discharge from the service, he came north and married Betsy McGrath. She worked as a teller in a local bank. They had no children.

I knocked on his door.

"Come on in," a voice said in a raspy Southern drawl.

Dan was sitting upright in bed wearing a light blue hospital gown. He had short jet-black hair, a closely trimmed salt-and-pepper beard, and a broad, thick nose that looked as if it had been broken more than once. An intravenous line was taped to his left hand. I noticed two framed photographs on his night table. One was of a group of young men in army fatigues standing around a jeep. The other was of his wedding day, with Dan in a tuxedo and Betsy in a floor-length gown. Now she sat next to his bed, a short, plump woman of Dan's age, wearing brown slacks and a tan blouse, her blond hair tied in a bun. Her eyes were bloodshot.

"Good morning," I said in a firm voice. "I'm Dr. Groopman. I'm a hematologist-oncologist."

"Dan Conrad, sir. Pleased to meet you." Dan offered his hand in greeting. I noticed how he tried to exert a strong grip.

"I'm Betsy, Dan's wife." I shook her hand as well. "What's a hematologist-oncologist?" she asked.

"It's a specialist in blood diseases and cancer. I'll be overseeing much of Mr. Conrad's care, now and in the future. I'm the guy who is going to work to cure him."

I spoke with purposeful determination, and I intentionally raised the prospect of cure from the outset; despite the massive size of the tumor and its aggressive characteristics, it could be cured. Paradoxically, lymphoma of this type was most responsive to chemotherapy: The toxic drugs work against dividing cells, so the more that grow, the more that get killed. Hematologists and oncologists were eager to treat patients like Dan with the effective new therapies.

But Dan's face did not brighten in response to my intentional bravado.

"I'd like to go over your medical history and then examine you," I said to Dan. "Will you excuse us?" I asked Betsy.

She leaned over and kissed her husband on the lips. "I love you, baby."

"I love you."

For at least part of the medical interview, I prefer to speak one-on-one with patients, because there may be issues or information that they feel more comfortable discussing privately. And in this case there were.

After confirming the background noted in his chart, I asked about drugs.

"I smoked a lot of grass over there," Dan said, glancing at the army photograph. "And God knows what other shit was in that stuff—excuse my language, Doc."

"No problem. I've heard that word before. I've even used it."

I said it was a possibility, but a remote one, that exposure to any potential carcinogenic contaminants so many years ago would have triggered a lymphoma now.

"But I never shot drugs. I hate needles. Couldn't stick them in me for love or money."

He'd had gonorrhea in Vietnam several times, and syphilis once. "I was an alley cat then, like most of the guys. Since I married Betsy, I haven't strayed, not one time. She's a good woman." He paused. "It'll be tough on her to be widowed so young."

"Who's being widowed?" I asked sharply.

Dan faced me. "I've been around death, Doc. In Vietnam and after. I feel it deep in me."

Dan's despondency stopped me cold. I was pumping both of us up for the fight, my bravado meant to energize us for the long road ahead. Perhaps I had sounded glib or unconvincing when I told him I was intent on curing him. "When I said to you and your wife that you can be cured, I meant it. It's not going to be simple, and it's not going to be easy. And there are no guarantees. But we have a lot of experience pulling a person through this. I'm a straight shooter. Whatever I tell you, good or bad, will be the truth. And the truth is that despite how widely the tumor has grown through your body, there is a real chance that we can get rid of it. Forever."

I had given this speech before to inspire the hope that we had a fighting chance to succeed. Usually, these words produced a tentative smile or calmed a patient's anxious eyes. Then he might ask for details, the plan of treatment, how long it would occur, how many drugs were to be given, how long radiation would last, what were the greatest risks, or exactly how many people were cured in his situation. I was prepared to answer each of the questions. But Dan's face remained inscrutable, and he said nothing.

I began the physical examination. There was a fading tattoo on his right upper arm of a shapely woman in a bathing suit. "I was young and stupid when I got this," Dan said.

"We were all stupid when we were young."

My findings on physical examination were the same as those recorded by the residents. I heard wheezes throughout his chest and felt a rock-hard mass in the center of his abdomen. But equally ominous was his chilling lack of spirit.

I summoned Betsy Conrad back into the room. She nervously smiled as she sat at the foot of her husband's bed.

Now was the time to outline the treatment plan. I took a moment to organize my thoughts, feeling somewhat off course in the wake of Dan's negativity.

"The good news is that the kind of lymphoma you have is very sensitive to treatment. Chemotherapy and radiation work best when cells are actively dividing, like they are in your case. We have four very strong drugs that, when combined, deliver a powerful blow against the cancer. We will radiate the mass in the chest to relieve the pressure on your airways. And I think we should consider adding another agent that is still experimental. It's called a 'monoclonal antibody,' a special kind of antibody that is made in the laboratory and targets the cancer cells. It works like a smart bomb, hitting the tumor head-on and largely sparing the healthy cells. By combining the antibody with the chemotherapy and radiation, we hope to increase the chances for cure."

As I spoke, I studied their faces. Betsy's held rapt anticipation. I was reaching her, but I was not reaching Dan.

"I know I'm throwing a lot at you all at once, and on the heels of the diagnosis. Let's stop for a minute and talk about what I've said so far before I go on."

Betsy looked at Dan. He maintained his silence. "You can't just cut it out?" Betsy asked me.

"Unfortunately, it's not possible with this kind of tumor. And in this case, the mass is around vital structures, like major blood vessels and the bronchi, the airways. It would be very risky to try to remove even a part of it."

I turned to Dan. "No one ever wants a problem like this, but given that it's happened, it's better to have gotten it now than ten years ago. Ten years ago we didn't have the kinds of drugs we do now, didn't have supportive therapies that reduce many of the side effects of chemotherapy, and didn't have the monoclonal antibody. You were a soldier. You know what it takes to win. Finally, we have the right ammunition on our side."

"Uh-huh," Dan said with little conviction.

I wondered if the suggestion of an experimental therapy was causing him concern. "Are you worried about the antibody? It's been tested with literally hundreds of patients, and it's safe, the only major side effect being an allergic reaction. We give it slowly, with other medications, to prevent that reaction. I don't want you to think we're offering this to you as a guinea pig."

"I'm not worried about that."

"I'll leave a consent form for you and Betsy to read that describes the experimental antibody, the rationale for adding it, and all the potential side effects. When I come back later, I'll go over it with you, and you can make a final decision about taking it. If you decide not to, that's okay, and we'll go with the chemo-therapy and radiation alone."

Next I focused on the four drugs, emphasizing that they were chosen after decades of study had shown them to be the best combination. The major risks were infection and bleeding. I then raised the issue of tumor lysis and how we would work to prevent it. "Again, we have a lot of experience with lymphomas like yours."

I left the Conrads and caught up with the residents. I outlined the strategy to treat intensively with the combination chemotherapy, radiation, and monoclonal antibody.

"Big guns," Virginia Chu said.

"The biggest we have," I answered, "short of bone marrow transplantation, which hopefully he won't need."

I paused, then asked whether any of the residents had sensed that Dan was depressed. "He seemed pretty negative in there, even when I tried to bolster his spirits," I said.

"We all picked up on that," Virginia said. "He must be terrified. I'll go back after rounds to ask if there's anything I can do to make him more comfortable. We're going to get a cot for his wife so she doesn't have spend another night in the chair."

I remarked to myself how things had changed since my residency. Now there was a genuine interest in a patient's emotional state, and a willingness to make time to accommodate it.

A glaring deficiency in medical education with regard to a patient's spirit at last had come to the foreground. Several factors had coalesced to cause this. During the 1980s, as an outgrowth of political activism in the prior decades, patients and their families became more vocal in expressing how they wanted to be treated. Part of this involved being informed, provided with details about a disease and its treatment that formerly had been restricted to physicians. With this knowledge, the patient and family could make choices more readily. This was termed "patient empower-ment," and along with the demand for medical information came the cry that the sufferer's soul was often neglected. The pressure on the medical establishment to address the spiritual dimension of disease came not only from patients. Physicians faced growing competition from alternative healers as patients flocked to them. These practitioners of non-Western treatments put traditional physicians on the defensive. Beyond the herbs and acupuncture and meditation prescribed by many of these alternative healers was the time they spent with patients inquiring deeply into their emotional state. They talked about treating the whole person, an approach captured under the rubric "holistic medicine." The cul-tural shift in America could not be ignored by those who guided clinical education, the deans of medical schools and the chairmen of clinical departments. At Harvard Medical School, for example, the curriculum was overhauled to create what was termed a "new pathway." It emphasized the patient's experience of illness, not just the biology of the disease.

Teaching sessions were devoted to communicating better with patients, reading an emotional state and responding to it appro-priately. In some medical schools, there were even courses on spirituality in medicine, with lectures delivered by clergy and patients themselves. There was much more that we in the med-ical establishment could improve upon, but at least a com-mitment had been made. The focus of rounds was broadening beyond the strictly physical considerations of clinical care.

Later that morning I returned with the consent form for the

experimental antibody. I handed it to Dan. He placed the paper on his night table without reading it. "I want to think things over some more," he said.

"The experimental part?"

"No. All of it."

I was not sure what he meant. "Are there questions I can answer? Maybe I didn't make clear how critical it is to begin immediately, even if we don't use the antibody."

For a long moment Dan did not speak. "Doc, I don't mean any disrespect, but it's just not worth it. I'm not gonna make it. I know it. Deep inside."

Tears welled in Betsy's eyes.

His words chilled me. I am not by nature superstitious, but I had learned over the years to pay close attention when patients said they felt they were going to die. One of the first patients I admitted to the hospital as an intern was a middle-aged woman with vague chest pains. After examining her, hearing normal heart sounds and finding clear lungs, and reviewing her chest X ray and cardiogram, both of which were normal, I told her that so far I had found nothing serious. Perhaps it was indigestion, reflux from her stomach irritating her esophagus. We'd try antacids. She gripped me tightly by the wrist and told me no, that it *was* serious, that she had this sense something terrible was going to happen, that she was going to die. I reassured her that such feelings were false, just a manifestation of the anxiety of being in a hospital. Several hours later, I was stat paged to her room. I ran there and found a group of doctors and nurses in the midst of an attempted resuscitation. She did not make it. At autopsy, we found an aneurysm of her aorta that had ruptured. The aneurysm had not been detected on my physical examination or any of the tests. At the time I wondered what had accounted for her sense of impending doom, whether the brain received neural input from the fraying wall of the aorta that somehow foretold the catastrophe.

"I know it's a shock," I said to Dan, "all of this happening at once, learning you have lymphoma, urgently needing treatment. But as I said before, and I'll say again, *you can be cured*. There's no guarantee, but we have a real shot. We can get you through this and return you to a full and normal life."

Dan shook his head. "There's . . ." He hesitated. "I know what it takes to battle. I don't have it in me. There's no point to it."

We were getting nowhere. I needed help. "Are you a religious person?" Perhaps the hospital chaplain could shore him up.

"Not really. Brought up Baptist. But most of the faith I had I lost in the army."

"Perhaps it would help to have the chaplain visit anyway?"

Dan shook his head.

"Then I'd like to ask Dr. Levin to see you. He's a liaison psychiatrist on the staff."

"I really don't need a shrink. I'm not crazy."

"It's not a question of crazy. He does interventional work, meaning that he helps patients through hard times. He's not going to talk to you about whether you loved your mother or anything like that. I'd really like you to meet him."

Dan grudgingly agreed but maintained that he didn't think talking with Dr. Levin would change his mind.

As I learned a few hours later, it didn't. "It's difficult to penetrate his resistance," Jon Levin told me. "He just keeps saying that he doesn't have it in him and there's no point in going ahead. Strictly speaking, he's not depressed. He doesn't show any slowness of thought or vegetative signs. I'll go by and see him again tomorrow, but I think it's going to take time to get to the core of why he refuses treatment."

Time was not something we had.

I visited Dan after Jon Levin's call. He was still working on his lunch. I noted how he took small bites, chewed them well, and waited for them to pass down. The mass was limiting peristalsis in his esophagus so he couldn't ingest usual volumes of food. Betsy was on the phone. She made her apologies to the caller and hung up. "Another of his army buddies calling to wish him well."

The blood tests done that morning showed that Dan's LDH was rising. LDH is an enzyme found in the blood that, in cases of lymphoma, roughly reflects the bulk and growth rate of the tumor.

"The blood tests show that the tumor is growing very quickly," I said. "It may soon strangle your airways, or block your bowel, or erode into your aorta."

Dan put his fork down. "Doc, I'm not the kinda guy who can be scared into changing his mind. I appreciate your intentions. Really, I do. But nothing that you say will move me."

Betsy looked at him plaintively and turned to me. "He is so

stubborn. Worse than a mule. I tried all day to convince him. He wouldn't even read the form you gave him on the antibody." She started to sob and reached for tissues from her purse.

I was at a loss. There was a real chance to eradicate his cancer. The cure of non-Hodgkin's lymphoma was one of the great triumphs of modern oncology. Dan's chances were just shy of fifty-fifty. When I started as a medical student, they would have been closer to zero. I had tried to give him hope, emphasizing that we had what it took to defeat his disease. And he was surrendering without a fight.

Since Esther Weinberg, I had encountered a few patients who initially declined treatment. In each instance, there was a stated reason. Some elderly patients felt the time that might be gained was offset by the diminished quality of life due to the therapy; thus they opted for less intensive therapy or waited until the disease caused symptoms that necessitated local palliative treatment. Such choices made sense, and I supported them. Other patients were unable to assimilate the information I gave them because of their fear. For them I needed to explain more clearly, or recruit a family member or friend to reinforce what I had said. Then there were people who believed they were undeserving of hope, like Esther. I didn't sense that was Dan's reason, although I thought it might emerge in discussions with Jon Levin. Not sure what more I could say to convince Dan of the necessity of starting therapy, I reiterated my point more forcefully.

"I want to be sure you understand what I am recommending. We should get started *immediately* with chemotherapy, four drugs used in combination, radiation, and, if you agree, the experimental antibody. We would treat until we reach what is called a complete remission, meaning that there is no detectable lymphoma. If it doesn't completely disappear with those four drugs, then we have many backup drugs, also used in combination. If, for some reason, there is still residual lymphoma after the backup drugs, then we would proceed to a bone marrow transplant. That hopefully won't be necessary, but we have it in reserve. We *can* cure you, Dan."

Betsy nodded in assent. But Dan was impassive.

I went on, "Again, think of it like a battle where you go in with air support and heavy artillery. You throw your strongest punches right from the start, beating back the enemy. If that works, then there is mopping up afterward, capturing or eliminating any hidden pockets of resistance. But if the enemy is dug in deeply, then you call in even bigger guns. At every point, we are here to support you, to carry you through, and we work like the dickens to bring you back healthy and safe."

When I finished, Dan repeated that his decision was made.

Jon Levin visited Dan before dinner. He spent nearly an hour with him, probing persistently, searching for some opening into why he would forgo any possibility to survive. But Jon again came up empty-handed.

I returned not long after. Virginia Chu, the resident, told me that Dan's oxygen level was falling. She had raised the flow rate, but it hardly increased the level. Dan was fatiguing. As he tired, the work of breathing would increase, and he might need to be placed on a respirator. Ordinarily, I already would have obtained an end-of-life directive; that discussion usually came on the heels of a commitment to treatment. Patients with a rapidly growing lymphoma were supported with all measures and placed on a respirator if necessary, since there was a chance of cure. Only if there was an irreversible event, like a cerebral hemorrhage that rendered the patient brain-dead, would we retreat from this intensive care.

To obtain an end-of-life directive now felt like sealing Dan's sense of impending death, reinforcing his decision. But I had no choice.

Dan was upright in bed, the oxygen mask over his face. His hair and beard were glowing with sweat, and his eyes were puffy. The TV news was on, but neither Dan nor Betsy was watching it. She eyed him tensely, while his gaze seemed to be on some indistinct point before him.

I drew a chair from the corner of the room. "I just spoke with Dr. Chu," I began. "Your oxygen level is hard to maintain."

Dan nodded knowingly.

"It's a terrible feeling, being unable to breathe."

Betsy's hand tightened around Dan's.

"Have you had time to reconsider treatment?"

"Doc, I thank you . . . and the young doctors . . . for what you all are trying to do," Dan said haltingly. "But I know . . . what I'm doing."

Betsy was fighting back tears.

"Dan, let's put aside for a minute the issue of treating or not treating the cancer. I am worried that you are fatiguing, that it's going to become increasingly difficult for you to breathe and maintain your oxygen level. We need to consider how to handle that. Because your lymphoma could be cured, we would want to support you, meaning put you on a respirator in the ICU."

Dan shook his head. My stomach tightened. "I know what that means," he said. "It's just prolonging the torture."

"Honey, please," Betsy pleaded, "please."

"I love you, baby," Dan said, "but I know . . . I'm right."

I had feared that this would be his answer. It put me in a very difficult position clinically. Usually, I would temper the anxiety and struggle that a patient experiences in respiratory distress with judicious amounts of a medication like morphine. But morphine could dull his drive to breathe and actually decrease his oxygen level. The goal was to shrink the cancer quickly enough so that the potential negative effects of the sedative would not be felt. It was a fine balance that could tip dangerously in the direction of a falling oxygen level if the tumor did not shrink fast enough and the airways did not open. In such cases, we would intubate the patient and place him on a respirator. The respirator was the safety net that assured adequate delivery of air until he was strong enough and less burdened with the cancer so as to safely breathe on his own. There was no safety net now. As Dan tired and his anxiety rose, his breathing would become less efficient. If I tried to relieve his anxiety with medication, I would likely decrease the respiratory drive from his brain, and his oxygen level would worsen, since the tumor would still be expanding.

"I don't understand what's going on here," I said, trying to control my frustration and my own fear.

Dan did not reply.

I sat for several moments but was met with silence. "The senior resident replacing Dr. Chu on call tonight is Dr. George Beckwith. He's also very good," I said. "I'm going to tell him that at least for now, you don't want to be intubated. But I'm also going to have him check on you like a hawk over the course of his shift, and he will contact me at any hour if you get worse."

Dan nodded his understanding.

I drove home, racking my brains to try to understand the impenetrable wall that Dan had put up. Jon Levin said that he would speak with him in the morning yet again. If Jon failed, then I would ask the chaplain to see Dan, even though he had said he was not a religious person. I just hoped that he would be strong enough to keep breathing until we were smart enough to find the roots of his resistance.

I called George Beckwith shortly before I went to bed.

"He's metastable," George said in the jargon of the house staff, meaning the situation was tenuous but unchanged. "I think another patient—someone not as physically strong, who doesn't work construction—would have tired long ago," he added. "The nurses are in there, and his wife knows to call if he starts to slip. Not much more we can do."

I slept fitfully, waking up every few hours and eyeing the red neon digits on my night-table clock. Finally, not long after four-thirty A.M., I got out of bed and made myself a strong cup of coffee. I had a busy day ahead of me. Later that morning, I would convene my weekly laboratory meeting, at which scientists presented new data from experiments conducted over the past days. It demanded total concentration, since it was easy to generate artifacts in experiments; the most minor details could trip you up and yield misleading information. A dynamic give-and-take marked these meetings. Researchers were taught to be point-blank and incisively critical, and to push the presenting scientist as hard as possible from every angle to be sure that the data are solid and set a solid stage for the next series of experiments. There were no arguments from authority, no bowing to statements made by more senior members of the group. That included me. When I presented my ideas and data, I would be challenged relentlessly, my assumptions questioned, my conclusions dissected. It was the only way to get to the truth.

I worried that my mind would be unfocused, my attention diverted by what was happening with Dan Conrad. A reason for his hopelessness was hidden somewhere, but all my efforts had failed to find it. And, unlike in the laboratory—where, when an experiment failed, you simply redid it more carefully, incorporating the constructive criticisms of your fellow scientists—at the bedside there were sometimes no second chances.

Shortly after the kids left for the school bus, my beeper went off. I was being paged to Dan's ward.

"Dr. Chu," the voice answered when I phoned in.

"Virginia, this is Dr. Groopman."

"Great," she said excitedly. "I just finished speaking with Mrs. Conrad. I think I know why Mr. Conrad has been refusing treatment."

My pulse quickened as Virginia Chu outlined the story. Minutes earlier, Betsy had taken yet another phone call from one of Dan's army buddies. Dan was too breathless to speak and told the caller so.

"Then the army buddy said to Mrs. Conrad that Dan's situation was just like the case of another veteran from their unit," Virginia related. Not long after the war, Betsy learned, one of Dan's closest friends had developed cancer. The caller did not say what kind of cancer, but Dan had kept a vigil at the bedside from the time his friend went into the hospital, through the multiple complications of the ICU stay, until the man died.

I said I was on my way in, that we should meet on the floor and talk with the Conrads. As I started the car, I thought of all the war metaphors I had used with Dan. If anything, they had backfired, only reinforcing his disabling memories.

I had failed to really understand Dan as a person, to probe deeply into the details of his life. It seemed easy to integrate his status as a veteran into the typical battle scenarios that I and other doctors painted for our patients. But instead of picturing himself as a victorious warrior, he must have seen himself as so much cannon fodder, like his friend. And I could well understand why. It had taken me years not to be jarred by what happened in an ICU, not to flinch at the bedside. I imagined the scenes Dan had witnessed, and how he must feel overwhelmed by the enormity of what he faced now. I had to do two things when I arrived at Dan's bedside: one, to sharply distinguish his case from that of his friend; and two, to somehow reduce in scope the prospect of the terrifying complications looming before him.

The traffic was still light as I entered the Longwood Medical Area. I parked, dropped off my briefcase in my office, put on my white coat, and made my way to the ward.

Virginia Chu was in her scrubs, sipping coffee from a Styrofoam cup. She put the coffee aside and knocked on Dan's door. "Come in," Betsy Conrad said.

The cot next to Dan's bed was still made up and had not been slept in. Betsy's face was drawn and pale. She stood holding Dan's hand. He nodded hello to us. I noticed how the muscles in his neck contracted with each inspiration. He was laboring as hard as he could to breathe.

I sat close to Dan, with Virginia to my right. "Dr. Chu told me briefly about your friend from the army," I said.

Dan closed his eyes and nodded.

"I don't have enough details about his case to comment on it clinically, but whatever happened, that doesn't have to happen to you." I paused, watching Dan continue to struggle under the oxygen mask. "I told you that I'm a straight shooter, that I would not lie to you. There is a very real chance that we can cure you."

Dan turned to me. His voice was muffled under the plastic mask, and the words came out on gasps. "The doctors . . . said . . . the same thing . . . to him."

"Maybe they were telling him the truth and there was a real

chance of cure. Maybe they weren't. I don't know. But you have to believe that there is for you."

Dan continued to gulp in air.

"He saw it all," Betsy said. "The effects of the chemotherapy, then when his buddy got pneumonia, they put him in the ICU, and he started to bleed and went into shock. They gave him all sorts of drugs, antibiotics, and transfusions. It went on for over a week, day and night. It was before we were married. Tom Kane. That's him." Betsy pointed to a soldier standing next to Dan in the photograph. He was a head taller than Dan, with sunglasses and a toothy smile.

"I'm sorry about your friend," I said to Dan. I conjured in my mind again the harrowing images of hemorrhage and sepsis and shock that could not be erased from Dan's eyes. I thought of patients like George Griffin who had found and sustained hope despite much longer odds than Dan; and of Esther Weinberg, who had not. At the core of their experience was a sense of being in control or at the mercy of forces outside of them.

"Let's go step by step," I said to Dan. "You don't have to make a commitment beyond taking the first step. Treatment is a train that can be stopped at any time. And you'll be the one in charge, the one to make the decision. I may not agree with that decision, but I won't be angry at you, and none of us will abandon you. Now I understand why you thought you didn't have it in you, and why you thought it was pointless. But you *do* have it in you. I can see it.

"So let's make a deal, with Betsy and Dr. Chu as witnesses.

Let's begin the therapy, the first course of chemotherapy and radiation. We can use the antibody or not, it's your call. And if you want to stop after that initial treatment, then we'll stop. If you choose to go on, then we go on. You are in the driver's seat all the way."

"Please, Dan . . ." Betsy said.

Dan's shoulders sagged, and he reached for a cloth to wipe a river of sweat off his cheek. I noticed how unsteady his hand was.

"Okay," he said.

I wrote the orders in Dan's chart. He signed the consent form for the monoclonal antibody. I had read it aloud and afterward asked him and Betsy if they had any questions. They didn't.

"Now we get into gear," I said to Virginia Chu.

Her face became animated. "I'll monitor him every few hours for tumor lysis syndrome," she said, "and of course watch his O_2 level. The radiation people already know about him. He'll need a good dose of antiemetics on board; with his respiratory status, vomiting could tip him over."

I agreed, impressed with how she had everything in place. "Keep me posted," I said.

Within hours, the four drugs and the antibody were infused and the radiation begun. Dan felt intensely nauseated but did not vomit. The pulmonary team was alerted to the possibility that he might need to be intubated and moved to the ICU. By evening his blood tests showed a fall in his serum bicarbonate and an upward drift in his potassium, signs of tumor lysis. Virginia Chu and the medical team delivered sodium bicarbonate and saline regularly, checking the pH of his blood to keep it in range. The electro-cardiogram did not show any change in heart rhythm.

"I'm rotating off now," Virginia said as we stood together in the hall. I thanked her for an excellent job. Another senior resident was taking her place for the next shift, and I asked him to keep me updated as needed.

I went in to say good night to the Conrads. Dan's face was flushed, and he was sweating profusely. I reassured him that this happened with high doses of corticosteroids, one of the medications he was given both to prevent nausea and to combat the lymphoma.

"Step by step," I said to him, and pressed his hand in emphasis.

I checked in with the senior resident at ten o'clock. No news, he said with relief.

The next morning I was greeted by a wide smile from Virginia Chu. "He looks much better," she said. "His bicarb and potassium are holding, and his calcium and magnesium are in range. The urine output was steady through the night, and his uric acid was up a little, but nothing worrisome." She handed me the computer printout with his laboratory values, and it confirmed what I had just heard.

"And he's breathing more easily. It's incredible how there can be this kind of change within twenty-four hours."

I explained that this was the case with such rapidly growing tumors. In addition to the mass's physical effects, it also caused inflammation and edema around the bronchi. The corticosteroids helped alleviate that.

I entered Dan's room. "Dr. Chu says you are a bit better this morning," I said.

He nodded. I noticed that, while his breathing was still labored, he was no longer using his neck muscles. Betsy offered a tentative smile as I drew the curtain around Dan's bed and examined him. His heartbeat was rapid and strong, and although there were still loud wheezes, the air was flowing better.

"The tumor is breaking down, but it seems to be under control, not disrupting your metabolism."

Dan said a muffled "Good."

"And your oxygen level is holding."

Betsy told me that for the first time in days Dan had been able to take in a few sips of broth. We had been feeding him intravenously, since his struggle to breathe had made it too difficult to eat by mouth.

"Another good sign," I said. I paused, not sure whether I should risk any departure from the positive atmosphere, but cases like Dan's could deteriorate unexpectedly and catastrophically. "You feel up to talking about how you want us to handle things should we get into a tough patch?" I was purposefully vague and tentative.

"I can change my mind later?" he asked.

He had spoken a full sentence without gasping.

"Like I said, it's a train you can get off of whenever you want."

Dan said that he did want to be supported and, if necessary, placed on a respirator and transferred to the ICU. Betsy's eyes grew moist as I said I would inform Dr. Chu and document his decision in his chart.

Dan's fever spiked to 104°F a week into the treatment. We had given him G-CSF, the protein that boosts white cells, anticipating the toxicity of the chemotherapy on his bone marrow. Despite that preventive measure, his white cells had fallen to a low enough level to make him vulnerable to life-threatening infection.

"I told you," Dan said. He was shaking from the temperature, and a nurse was working to bring it down by sponging him. "I told you. Just like Tom Kane."

I was struck by how quickly Dan fell back into despondency. It was hard to predict all the complications that might ensue, and many could mimic what had happened to his friend. I feared he might prematurely call for a stop to his therapy.

"Each person is different," I said. "You are Dan Conrad, not Tom Kane. Lymphoma is not lung cancer. Always keep that in mind. Despite setbacks, the bottom line is that you can be cured. We'll treat this presumed infection with broad-spectrum antibiotics, and we will support you until your white cells come back."

I paused. Dan's eyes communicated deep doubt. "Step by step, Dan. We took the first step, and the tumor is shrinking. This is the second step, to move you through this. We can. I want you in the ICU."

The nurse's cooling sponge had caused his teeth to chatter. Dan nodded his shaking head, which I took as a statement of affirmation.

We isolated a gram-negative bacterium in the blood cultures, a common infection in people whose immune systems are low.

But in Dan's case, since such bacteria often come from the bowel, I worried that the tentacles of lymphoma in his abdomen, breaking down with the treatment, might erode the wall of the intestine. If the intestine perforated, it would precipitate peritonitis and shock.

The next morning Dan complained to the ICU resident of pain in his lower abdomen that felt like a knife slicing through him. "I've been cut before," Dan said, "and that's what it feels like."

A surgeon was called and X rays ordered.

"I think he needs to be explored," the surgeon said to me. "We don't see anything definitive, but if we wait for it to declare itself, it may be too late."

I went to meet Dan and Betsy in the ICU. Dan had received some low doses of morphine, but I could still see the struggle playing over his face as he tried to endure the pain.

"I know you are really uncomfortable," I said. I grasped Dan's hand. He closed his eyes and said nothing. "And it's hard to see what's at the end. But nothing has happened that we can't fix. I mean that."

"The surgeon said they may have to cut out some of his bowel," Betsy said.

"That's true. We don't know exactly why Dan is having so much pain. The lymphoma is breaking down. We know that from the scans. It's shrinking. As it breaks down, it could damage normal structures. It could have penetrated into some organ, like the bowel, or into the wall of a blood vessel. But the treatment is working.

"We made a lot of progress with the tumor in your chest," I said to Dan.

"I know," Dan said weakly. "I can feel it. My breathing is easier. The resident said I don't need as much oxygen coming through the mask."

"And you will feel a lot better when we deal with whatever is going on in your abdomen. The surgery itself is no picnic, with general anesthesia and the recovery from the operation. But we should just keep slogging through."

"I've been here before," Dan said with a faint smile, "deep in the mud."

He had taken the first step, the drugs and the radiation. That was the hardest. The second step was easier.

The operation lasted nearly six hours. The surgeon found what is called "ischemic bowel," meaning that one of the arteries to a segment of intestine was obstructed by clot due to a thick tentacle of lymphoma wrapped around it. If the segment of bowel was not removed, it would die and, in dying, perforate and release the bacteria that filled its content into the abdomen. Dan would have gone into shock and likely died from such an event.

We were fortunate that the surgery occurred when Dan's blood counts were on the upswing. He had an expected post-operative fever, but cultures done from his blood, urine, and sputum were negative for bacteria.

"Optimally," the surgeon said, "I'd like to give him another week or so to let the wound heal."

We were discussing the need for further chemotherapy. The drugs acted against rapidly dividing cells. To heal a wound, cells grow more quickly than usual, forming a bridge and laying down scar. The chemotherapy might affect these cells as well, causing the reconnected bowel to come apart at the suture line.

"It's a tough call," I told the ICU residents. "If we wait too long, the lymphoma may start growing again and get out of hand. If we treat too soon, Mr. Conrad will be back in the OR with a leaking bowel, causing peritonitis."

I explained this to Dan and Betsy.

"I feel like I'm going backward rather than forward," Dan said.

"No," I replied. "Think of it as stepping to the side for a moment. The only step backward is if the tumor were to grow despite the treatment. We haven't lost any ground. After the surgery, we'll get back on track."

To give the wound a better chance to heal, we waited a few days after what would have been the date for the next course of treatment. Then the four drugs were administered. I was reluctant to cut back on the doses of chemotherapy, unwilling to give the tumor any respite. As the orders were written, I went over a checklist with the residents of possible complications during this perilous period: a drop in the white count and another burst of bacteria into his bloodsteam; perforation of the bowel at the site of the sutures; erosion of the tumor into a blood vessel with hemorrhage.

At the end of each day, I felt a quiet sense of relief that no catastrophe had occurred. As two weeks passed and the white blood cells rose again, it seemed that the worst was over. "You are coming around well," I said to Dan. "So well that we can transfer you out of the ICU and back to the floor."

"I feel better," Dan said. "My breathing is easier, and I can eat without feeling like things are getting stuck."

After ten days Dan was due to be discharged from the hospital. The mass in his chest had become a faint shadow, and the tumor in his abdomen had shrunk by over 50 percent. The nurses had given him his discharge orders, reviewing with Betsy exactly what medications needed to be taken each day, and what side effects to look out for. He was scheduled to return in two weeks for his next course of treatment.

It was early March, about ten weeks since Dan came to the ER, and a light snow blanketed Boston, giving the city a sheen of pristine beauty. I entered the room just after the Conrads had finished packing their suitcase. Dan's nurse caught my eye and raised her eyebrows, but I was not sure what she was signaling. Then I looked at Dan. Instead of a face of celebration, I saw tense glumness.

"Everything okay?" I asked.

"Sure," Dan said.

Betsy occupied herself by scanning the room for unpacked items.

"Breathing still good and digestion back where it should be?"

"Yup."

I thought, then returned to the issue of Tom Kane. "Did your friend have some problem when he left the hospital?"

"No. Remember? He never left the hospital."

I said I was sorry. It had slipped my mind that he died in the ICU.

"Then what's the matter?" I feared his natural reticence and how easily I accepted it. It had been only through serendipity that Betsy learned why Dan had refused therapy. "We've been through a lot these past weeks. You're not going to hold back on me now."

"I just . . . I'm just scared. I'm scared to leave the hospital."

"That's normal," I said. "If you weren't scared, I would call Dr. Levin back to see you. But it's not like we're cutting you loose. You are still tied to us; we're stretching the umbilical cord a little longer. You and Betsy know that you can call any hour of the day or night. If for any reason I'm not available, then the hematology-oncology fellow will be paged. If it's an emergency, then come straight to the hospital and don't even bother calling."

Dan nodded unconvincingly.

"He's got in his head again that he's not going to make it," Betsy said.

Dan looked at her fiercely.

"Why?" I asked.

"I don't know."

"Aren't you feeling better, or do you feel something wrong inside?"

No, he replied, he was feeling much better.

"I told you there were no guarantees, but every day that passes brings us one day closer. It's a long, long haul, and we're all with you through it."

Dan bit his lip nervously. There seemed nothing more to say. But the image of his downcast face lingered with me into the day.

———

Dan no longer needed to be admitted for his treatment; he could receive it safely as an outpatient in the chemotherapy clinic. I met him there two weeks later. A strong wind blew from the north, bending the leafless trees.

Dan shed his parka and handed it to Betsy. He shifted on his feet. The three of us walked into the treatment room. Chemotherapy is administered in a large, open space in a special area adjacent to the outpatient clinic. There are twenty or more people at any one time receiving therapy. Some sit in chairs that resemble a La-Z-Boy; others recline. Nearly all have lost their hair. Some sport baseball caps or bandanas, while others wear wigs. That day Dan was wearing an olive-green cap.

"Army issue?" I asked.

"No. Just something cheap that I picked up at a flea market a year ago."

As I escorted him and Betsy to the far end of the room, we passed people with sunken cheeks and hollowed eyes. It was impossible not to think that some of them would die.

The nurse attending to Dan that day was Deirdre Dolan. She was one of the first chemotherapy nurses I worked with when I returned to Boston from UCLA. She was efficient, no-nonsense, and every bit as knowledgeable as any attending oncologist.

I introduced the Conrads to Deirdre, and she gave them a welcoming smile. Dan's face tightened. "It's going to be okay," Deirdre said softly. He didn't reply.

I pointed Dan to the upholstered chair where he would sit, but Deirdre interrupted. "Mr. Conrad, I don't want you in that corner there. Here," she said, taking his hand, "come over to this chair."

Deirdre led Dan several stations away, with Betsy in tow. I wasn't sure why Deirdre wanted Dan there, but I assumed that since she was attending to several patients simultaneously, she wanted to cluster them for convenience.

I wrote the orders for the chemotherapy and monoclonal antibody, then told the Conrads that I would check in with them a few hours later when the infusions were finished.

It was nearing lunchtime when I approached Dan and Betsy. They were snacking on graham crackers and juice, standard fare in the chemotherapy clinic.

"Want one?" Dan asked me.

"Sure."

As I chewed the cracker, Dan said good-bye to the woman in the adjacent chair. She appeared to be in her early sixties and wore a Boston Red Sox cap. Strains of fuzzy gray hair peeked out from the back. I didn't recognize her, but that was not unusual, since several hundred patients passed through the treatment area in any given week.

"She was in to have her blood counts checked and receive a transfusion," Betsy said.

"Nice lady," Dan added. There was a distinct change in his demeanor.

"Deirdre said the treatment went without a hitch," I said to Dan.

"It did."

"It must be a relief to come in and be treated without having to be in the hospital."

"It'll be a real relief when I'm like that lady."

I looked quizzically at Dan, and he laughed.

"She ended treatment two months ago, also for lymphoma," he said. She had told Dan that she had her rocky times as well, but she pulled through.

Deirdre Dolan came and deftly removed the infusion needle from Dan's arm. "Keep some pressure on that for a while," she said. She turned to me. "You don't know Dotty Hirschberg?"

I shook my head.

"She's a patient of Dr. DeNucci's. She was also on the antibody protocol."

I caught the glint in Deirdre's eye.

Dan completed his course of treatment, the only complications moderate fatigue and anemia. We treated him with erythropoietin, which increased his red blood cell count, but toward the end of the multiple cycles of chemotherapy, he required transfusions. There were no further episodes of infection.

"You are a free man," I said to Dan when the last CAT scan showed no evidence of tumor.

"Not really. You said five years."

I replied that he was right. The rule of thumb is that cure is certain only after five years have passed.

"When do you think I'll get back to work?"

I told him I wasn't sure. Often there were several months of fatigue after intensive chemotherapy. It was important for him to build himself back up by eating well and beginning to exercise. I would see him every month for a year to monitor his progress, and then we would space out his visits after that.

It has been nearly a decade since Dan's treatment ended. He is cured. When I see him on his yearly visit, I leave the clinic feeling lighter.

The experimental antibody that Dan received was further modified by genetic engineering, and along with several related antibodies, these agents have emerged as mainstays of therapy for non-Hodgkin's lymphoma. The rate of cure for this cancer continues to rise with refinements in combining different modalities like chemotherapy, monoclonal antibodies, and radiation. Dan was one of the first to receive such combination treatment.

It was Dan who made me acutely aware that clinical medicine is a series of overlapping puzzles. Doctors are taught in medical school, residency, and fellowship that they are detectives in search of clues. These clues may be found in the patient's past medical history, whether he had certain illnesses as a child or more recently; or they may be uncovered in the social history, whether the patient drinks or smokes or comes in contact with a pathogen or carcinogen at the workplace. The family history is also important, since there is a growing awareness of genetic pre-disposition to disease. Of course, the physical examination and laboratory tests, X rays and scans, all can point to the answer. But none of those diagnostic modalities reveals clues to the second puzzle: Why does a person lack hope?

It was by chance that Betsy Conrad heard the story revealing the reasons for her husband's resistance. Tom Kane embodied for Dan a nightmare that he believed he would soon enter, one that would end in death. We all seek models of hope and despair, and our sense of hope or despair is reinforced by direct contact with someone who has either prevailed or perished. But more impor-tant, to help a person find hope, you need to know him in real depth. Within his past are the clues to the second puzzle.

Dan's sense that he would die was articulated in vague and vis-ceral terms. From seeing cases like his and that of the woman with the aortic aneurysm who felt impending doom, I've come to believe that the way the body talks to the brain powerfully shapes our sense of hope or despair. Dan's feeling of growing suf-focation, the increasing pressure on the esophagus and bowel, the compression of blood flow in the aorta and vena cava, I suspect, were all subconsciously perceived in his brain and translated as prefiguring death. The cerebral processing of that visceral input as a signal of death was accurate. Without the kinds of therapy that had been developed over the decades, this cancer would have been fatal. Hope, then, is constructed not just from rational deliberation, from the conscious weighing of information; it arises as an amalgam of thought and feeling, the feelings created in part by neural input from the organs and tissues.

That is why, I have hypothesized, it became easier for Dan to affirm hope within hours of beginning treatment, because his vital life functions—respiration, digestion, and circulation—all improved. There was a change in messages deep in his brain; it was receiving signals of growing health rather than of looming death. The steady progress made in unburdening his lungs and vessels and bowel nourished his hope.

Hope can be imagined as a domino effect, a chain reaction, each increment making the next increase more feasible. The familiar poetic phrase "hope flowers" captures this catalytic process. And the growth of hope is not strictly linear, always expanding. There are moments of fear and doubt that can deflate it, as when Dan left the hospital or entered the chemotherapy clinic. Deirdre Dolan, the oncology nurse, knew that more than words was needed. She purposefully positioned Dan next to Dotty Hirschberg and, by so doing, offered a model of hope to combat his despair. Dotty was living proof that a cure was possible. Her face finally replaced Tom Kane's in Dan's mind.

CHAPTER 5

Undying Hope

After decades of practice and seeing many hundreds of patients with a particular disorder, a physician becomes very familiar with the various clinical aspects of the malady, yet he never stops learning about character. This is one of the great privileges of doctoring, because what you learn can inform how you want to live your own life in addition to your care of future patients.

In December 2000 I met a patient named Barbara Wilson. Barbara was sixty-seven years old, a retired history teacher, active as a volunteer in her church and the leader of its Sunday-school classes. Her husband had died suddenly in his forties from a ruptured cerebral aneurysm. They had no children but many nieces and nephews, and Barbara was close to them. Some three years earlier, she had found a lump in her left breast that was biopsied and shown to be cancer. The dogma that a radical mastectomy, the operation Esther Weinberg underwent, was optimal therapy had been shown as misconceived. Such drastic surgery offered no clinical advantage over less mutilating approaches. Barbara's surgeon detailed two equivalent procedures: a "simple" mastectomy, in which the breast tissue and nearby lymph nodes would be removed and the underlying muscle preserved; or a "lumpectomy," in which only the tumor and surrounding lymph node would be excised, followed by radiation to the breast. Barbara chose the latter option.

The pathologist examined the excised cancer and noted that it had aggressive features, with many actively dividing cells. In addition, the lymph node contained metastatic deposits. Barbara was prescribed six months of chemotherapy, the standard amount to try to eradicate any residual cells and improve the chances that the cancer would never return. But a month before we met, Barbara noticed a hard swelling over one of her lower ribs. She went to her local oncologist. He examined her and ordered a series of X rays and scans. The cancer had returned. It was growing in her bones and liver. Her oncologist was in the process of retiring after many years of practice, and he referred Barbara to me for continuing care.

It was a typically frigid December day in New England, the winter sun reflecting sharply off the ice that had lingered on the sidewalks since the last storm. Barbara was seated in the waiting room, wearing a heavy woolen sweater, flanked by two women friends from her church. She was tall and slim, with angular features and serious blue eyes. After introductions were made, I asked whether she wanted her friends to sit in on our discussion. Generally, I encouraged patients to have a third party, a family member or friend, along for the part of the conversation when difficult clinical issues were aired. It facilitated communication, because fear or anxiety often made it difficult for a patient to assimilate information or summon questions.

"Do I need bodyguards?" Barbara asked with a laugh. "I suspect what I am going to hear will unnerve them more than me. For now let's leave them out here with last month's *Time* and *Newsweek*."

Barbara entered my office and surveyed its walls with a discerning eye. She lingered on a pastel drawing of a rainbow; below the rainbow were verses from Genesis in which God promises Noah that after the Flood, life on earth will never again be threatened with extinction.

"God has kept his promise," Barbara said, "despite what some men have done." She sat down across from me, her hands folded in her lap.

"I know you were told the results of the biopsy and scans," I began, "but let's review them anyway."

Barbara nodded. There wasn't any sign of distress or fear on her face.

"The bone scan showed deposits of tumor in the spine, pelvis, and ribs," I said. "Plain X rays indicate that none of the bones is at risk to fracture, so we don't need to pin or radiate."

"Well, that's a relief."

"There are three distinct tumors in the liver on the CAT scan, each measuring about four to five centimeters, or two inches. One piece of good news is that your brain is negative."

Barbara smiled. "I expected as much, given how little there seems to be in there."

I laughed, but not wholeheartedly. While humor can be an important coping mechanism for some patients, Barbara's poised demeanor raised my wariness. I continued, "The cancer cells have an aggressive appearance under the microscope, meaning that they are likely to grow and spread further. I believe we should treat the tumor aggressively—"

"So do I," Barbara interjected. "I want to make it plain that I wish to live as long as possible, so long as there is quality to my life."

Barbara had jumped one step ahead of me. Usually, defining what limits might be placed on treatment came after a lengthier discussion of the options and considerable reflection by the patient. Barbara obviously had arrived prepared to set the parameters. But I still needed to make sure that she understood what "aggressive treatment" meant in terms of side effects and risks.

"The tests done on the biopsy tissue show that the breast-cancer cells have what is called HER-2 protein on their surface. This means that we can go after them with standard chemotherapy and with a new treatment called Herceptin. Herceptin is an antibody that targets the tumor and increases the chances for a remission. But we have to be very careful, because you've had chemotherapy before, and combining the new chemotherapy with Herceptin has potential side effects, particularly damage to the heart muscle. This damage can result in heart failure. We will monitor you very closely, following your cardiac function with scans to avoid this complication. There are also the usual side effects of chemotherapy, lowering blood counts so that you may become anemic, or making you susceptible to infection or bleeding. We have good supportive therapies, like erythropoietin and Neupogen, that help boost your red cells and white cells in case they fall."

It was a lot of information at once, but Barbara affirmed that she understood what I said, because she was a "veteran" from her prior treatment.

"Let's look at the best-case scenario," I said. "The treatment causes the cancer to regress, and you enter into a remission. Remissions can last for months to many years. Hopefully, yours will be a very, very long one. During that time you can live a full and active life, working at the church, traveling, enjoying your family, whatever. That's the goal."

I paused. At this point, many patients asked for more details, particularly their chances of going into remission and the average length of remission. Barbara didn't. I wasn't sure whether she didn't want to know or had already learned the answers from her former oncologist or books or the Internet. But I needed to inform her of the uncertainty and limitations of the treatment as a segue to difficult issues that were best defined from the outset.

"It's important to understand that at this point in time, strictly speaking, even if we achieve a remission, there is no cure for breast cancer once it has metastasized. Treatment is palliative."

"I know," Barbara said. Her expression remained placid.

I paused, but she said nothing more.

"I have had women live for years with breast cancer, like it was a parasite," I said, purposefully invoking a stark image, "and each time it rears its head, we knock it down again with another treatment. As you know, there is a tremendous amount of research to find new ways to attack breast cancer, to develop more effective and hopefully more definitive therapies. That's the hope we should always keep in mind."

"I don't mind being a guinea pig, if we reach that point," Barbara said.

Again she seemed to be ahead of me. She had already anticipated the issues that I typically unfolded in sequence. So I went on to the bleaker scenario.

"It's important to talk about what we should do if things don't go well, if the treatment doesn't shrink the tumor or works for only a short time. We have to consider that the cancer can spread to other parts of the body, like the central nervous system, where the therapies we have are very limited in effect."

"Doctor, let me give you this." Barbara dug into a canvas tote bag. She took out a book, a thick biography of Churchill, and then a small, well-worn pocket-size Bible, and placed them on her lap while still digging in the bag. "Here it is." She handed me a sheaf of papers. One was a notarized health-care proxy, indicating that her minister, William Babcock, was authorized to make decisions should she become incompetent. The other was an end-of-life directive stating explicitly that should she lapse into coma, or should the quality of her life deteriorate significantly, no excessive measures were to be taken, such as placing her on a respirator or giving her cardiac resuscitation.

I inserted the documents into Barbara's chart.

"I've lived long enough to know that things are not predictable," Barbara said. "My husband was never given the opportunity to make choices."

I noted once more Barbara's calm and poise and wondered whether they were real. She showed no fear, no anxiety, despite the prospects we were contemplating. I told myself that I needed to monitor her outlook as closely as her heart function and her blood counts. Like George Griffin, she gave no hint of denial about her condition or her prognosis. Could someone really transcend the deep fear of death that seems to mark us all? If so, I wanted to know how.

Barbara's chemotherapy was scheduled to begin at the end of the week. She arrived alone, saying a friend would pick her up and drive her home. She walked easily into the treatment room, occasionally catching the gaze of a fellow patient receiving an infusion of drugs or in the midst of a blood transfusion. She smiled at each and, if the fellow patient returned her gesture, offered a gentle "Good morning."

"I am on the healing committee of my church," Barbara said. She explained that this group assisted the sick in the congregation, whether at home or in hospital. "When you've been a patient yourself," she continued, "you realize how much more it means for someone to show up in person rather than send one more vase of flowers or another Hallmark card."

I said I fully agreed, but not everyone had the fortitude to enter the room of a sick person, and some people felt awkward being so close to illness, not knowing what to say.

"I'm not judging those who send flowers or cards instead of visiting," she said. "Just that being present, even if you feel awkward, is so comforting."

Barbara had the expected side effects from the treatment. She lost all her hair, but despite the suggestions of her friends, she refused to buy a wig. Instead, she sported a series of bandanas that she tied tightly around her head.

"All I need is a Harley-Davidson, a pair of aviator sunglasses, and a beer, and I'll be set for the part," Barbara said when I commented on one bright red bandana.

"You forgot the tattoo on your biceps," I said.

"How do you know I don't have one?"

"Because I examine you regularly."

"Well, one day I may surprise you."

Three months into the treatment, Barbara was entering a remission. The metastases in her bones and liver shrank to half their size, and no new deposits appeared. Scans showed her cardiac function was stable. Still, the therapy took its toll. For days after the treatment, Barbara was too exhausted to work at the church or teach Sunday school. But as these acute effects waned, she was able to assist Reverend Babcock part-time and to read the weekly lesson to the children. When the effects of the drugs were at their least, in the days before the next course, she usually traveled to see friends outside the Boston area. Barbara was a Vassar graduate, and one weekend several of her classmates arranged theater and dining in Manhattan.

"But you ruined my taste buds!" Barbara said to me when she returned from New York. A common side effect of chemotherapy is a change in gustatory perception. "I wangle a free dinner at an extravagantly expensive restaurant, and everything tastes like tapioca." Her tone turned more serious. "I saw the books in your office, and the lithograph of the rainbow," she said. "You are a person of faith. And I heard that sometimes you pray for your patients."

I looked down sheepishly. Everything Barbara said was true, but I'd never had a serious discussion about my religious beliefs with a patient.

Barbara sensed my uneasiness. "Do you know the joke of Saint Peter at the Pearly Gates?" she asked.

I didn't.

"Well, you know that Saint Peter stands at the Pearly Gates, and he's the one who admits you to heaven."

"I know that much, at least."

"Good. Well"—Barbara's face lit up as she retrieved the joke in her head—"one day there's a particularly large number of saints waiting to enter heaven. They're all standing in line. Saint Peter, of course, has to check their credentials one by one, and that takes hours." Barbara paused. "But, these being saintly people, they all wait very, very patiently. Then, all of a sudden, out of the blue, someone wearing a white coat with a stethoscope in his pocket jumps the line, walks to the front of the queue, and breezes in through the Pearly Gates.

"Now, even though these are all saintly people, they've been waiting a very long time, and understandably, they get pissed off." Barbara's smile widened. "So, anyway, one of the saintly people approaches Saint Peter. He says: 'Saint Peter, we've waited a long, long time, very patiently, and then all of a sudden that guy wearing a white coat with a stethoscope in his pocket just walks past you right into heaven. What's that about?'

"'Oh,'" Saint Peter replies. 'That's God. He thinks he's a doctor.'"

Barbara's eyes lit up with impish delight. But I knew that like a doctor speaking to a patient, she had used her words very carefully.

Barbara's remission lasted seven months. Then she noticed vague discomfort below her right ribs and decreased appetite. I ordered blood tests and a CAT scan. The tumor had started growing again in her liver.

"More rabbits to pull out of the hat?" Barbara asked when I told her the results of the scan.

I said there were.

"How good are they?"

I explained that after the first-line therapy, we had backup drugs. We should stay hopeful, but in general, the chance of remission was lower, and the duration of remission was usually shorter than before.

Barbara took in this information with a pensive look. "You know what my limits are," she said, "and I think you know what my aims are. I want as much time as possible to live as well as I can. I have many things that I want to do and can do. I'm not ready to give up."

I said I wasn't, either. We would go into this with our eyes open. "I want to start with Xeloda," I said. "It's a pill, much easier to take than the intravenous chemotherapy."

"But I don't want you to give me something just because it's easy," Barbara said. "I want something that has the best chance of helping."

"It's the best second-line drug after what you had. And I hear you."

Barbara had the expected side effects of Xeloda, inflammation of her hands and feet. I had emphasized that she should inform me at the onset of these symptoms, but her stoical nature was such that I learned about it only when the inflammation was in full bloom.

"It is a miserable bore," Barbara said after finally acquiescing and taking pain medication. "Xeloda is easier, you said. I can't type Reverend Babcock's correspondence." Her palms and fingers were beet red and cracked.

I told her I understood her frustration. The other drugs had their own side effects, particularly on her blood counts, and were likely to be more toxic after all her prior therapies. That was why I had chosen Xeloda. The reaction to it would pass.

She said, "The worst part is not being privy to all the gossip that I glean from Reverend Babcock's letters. I can put up with the pain, but to be denied the pleasure, that makes it unbearable."

I returned Barbara's smile, but my heart ached. She was putting up a brave front. "I'd like to give you a prescription for a stronger painkiller."

Barbara paused and then agreed.

"I don't believe we get extra points for unnecessary suffering," I said.

"The key word in your statement is 'unnecessary,'" Barbara said as she dropped the prescription in her tote bag.

After Barbara left, I sat in my office. My eyes moved to a book on my shelf that I had read not long before: *Doctoring,* by Dr. Eric Cassell. Cassell, a primary-care physician, wrote it at the end of a long career. He articulated a feeling I had vaguely sensed but never fully realized. There are some patients whom a doctor grows to love. It is a unique type of love, distinct from any other type of love the doctor has experienced before. It moves outside the bounds of the usual doctor-patient relationship; feelings and thoughts are no longer strictly professional and are shared among true friends. Barbara had sparked that love in me.

"It's working," I said excitedly to Barbara some ten weeks after she began the Xeloda. I pointed to the large round, lucent circles on the earlier CAT scan and then to the smaller circles on the repeat scan. "The metastases in the liver are not down to quite half their original size, but they're on the way. It's nearly a partial remission."

Barbara's eyes grew moist.

"Any improvement in your hands?" We had been forced to reduce the dose of the drug because the inflammation flowered into widespread blistering. Barbara had resisted, but I explained that persisting at full strength risked severe damage to the skin that was not easily reversed.

"They quieted down. But they're still awful, especially at night. I need two Percocets to sleep," Barbara said. Then she smiled slyly. "I figured out a solution to get all the delicious gossip." She lowered her voice to a conspiratorial whisper. "I ask different women in the congregation to pitch in and help Reverend Babcock, since I can't type with these hands. Then, under the guise of checking with them about how the work is going, I pump them." She paused dramatically. "Actually, you don't have to pump them too hard. They blurt it out."

Xeloda kept the cancer at bay for another two months, and then, as happens, it waned in effect. Mutations regularly occur when cancer cells divide, and over time, certain of these mutations make the tumor cells resistant to the particular drug.

The many months of nonstop treatment had worn Barbara down. But when I asked her if she wanted to take a break or stop entirely, she said that she felt the time had not arrived. "There are many moments during the day that still give me pleasure," she said. "Let's keep going."

Her case differed clinically from George's. His cure had been, as he said, miraculous, and it had also been steady, meaning that with each intervention—chemotherapy, radiation, surgery, and then more chemotherapy and radiation—the tumor had progressively given way. There was no evidence that it ever became resistant to the treatments. Barbara's cancer had, as is usually the case. George had shown me that one should always begin with hope. I couldn't see, with this last reversal, how Barbara would be able to sustain it.

Barbara received another three courses of chemotherapy, but the tumor seemed to shrug off the drugs. The deposits grew in her liver and in her bones. She lost weight and spent most of the time in bed. After the last cycle of chemotherapy, I admitted her to the hospital with a high fever. Antibiotics stemmed an early bacterial infection.

As Barbara slowly recovered from the infection, I told myself I knew of no drugs, either standard or experimental, that stood a real chance of ameliorating her condition. The time had come for me to tell her.

I chose to visit in the early evening, when the hubbub of the hospital had settled down, so there would be less chance of distraction and interruption. Barbara greeted me warmly, as she always did. I moved a chair close to the bedside and grasped her hand. She returned the gesture, but it had little force. After we chatted for a short time about several articles in the day's newspaper, I began to break the bad news.

"Barbara, we've known each other for well over a year, and we've been honest with each other every step of the way."

Briefly, her lips trembled, and then she regained her composure. Her eyes told me she knew what I was about to say.

"I know of no medicines that I can give at this point to help you."

We sat in heavy silence.

Barbara shook her head. "No, Jerry," she said. "You do have something to give. You have the medicine of friendship."

The next day when I visited, Barbara's minister was there. Although he had visited many times before, I had not crossed paths with him. A lanky man with thick white hair and alert green eyes, he wore a clerical collar and a plain jacket. He stood and introduced himself as Bill Babcock and then began to excuse himself.

"Stay," Barbara said in a voice that left no room for disagreement. "Dr. Groopman is here on a social visit, like you."

"I can come back later," I said.

"Three's hardly a crowd," Barbara said in the same definitive tone. "Anyway, we weren't discussing state secrets—just the hymns I want sung at my funeral."

Her voice was even. She could have said they were discussing the menu for next week's church supper.

I sat down, as did Reverend Babcock. He recommended several parishioners who could play the organ at the funeral service, and Barbara picked one.

"The eulogies must be limited to five minutes, not a minute more," Barbara said. "If there is one thing I cannot abide, it is those endless, droning eulogies, heaping on praise and not mentioning the juicy sins everyone really wants to hear."

Reverend Babcock shook his head in mock dismay.

"And don't think," Barbara said, turning to me, "that I don't have a trove of peccadilloes."

"No more than anyone else, Barbara," Reverend Babcock said.

"I have enough. Pride, above all. And, of course, moments of terrible envy. That Tenth Commandment, Bill, shouldn't be a commandment at all. It's impossible, just impossible."

For months she had sustained a determined spirit. Now that she knew the grim reality of her condition, I had expected to see a change. But she seemed undeterred. I wondered if it would prove to be a facade that would ultimately crack. Or was it actually possible to subsume fear and face death with such apparent equanimity? The idea left me dumbfounded.

Each time I visited Barbara, she had company. Mostly, her visitors were congregants from her church; on two occasions she saw her niece, Anne, now living in Rhode Island, and Anne's young baby. I made myself an observer, listening to Barbara's banter.

One Monday evening, the sun not yet set, I entered Barbara's room and found her, for the first time, alone. The five o'clock TV news had just ended.

"Mr. Putin is a tough one to figure out," Barbara said. She

looked at me playfully. "Have you come to give me my evening dose of friendship?"

I said yes, but then I retreated to clinical concerns. "Are you having much pain? Truth, now."

Barbara's face fell. "Always the truth, from the alpha to the omega. The nurses are very attentive. There are moments when I get really severe pains here," she said, moving her right hand to her lower ribs, over her liver. "The extra morphine shots seem to do the trick."

"We'll increase your regular dose to avoid these breakthrough episodes."

"But not so much that I'm nodding off," Barbara said firmly. "I want to be alert as long as possible."

I assured her we would adjust the dose to alleviate the pain but not be overly sedating. Barbara seemed satisfied with this.

The last rays of the sun were waning, and Barbara turned on the lamp over her bed. Her eyes were sunken, and her skin was pale. It would not be long, I thought. I knew how much I would miss her.

"Are you afraid?" I asked. Her months of candor had encouraged mine.

"You know, not really, not as much as I thought I might be."

I moved my chair closer to her bed. "Why do you think that is?"

"I'm not entirely sure. I have strange comforting thoughts." Barbara shifted onto her side so she could face me. She winced until she found a comfortable position. "When fear starts to creep up on me, I conjure the idea that millions and millions of people have passed away before me, and millions more will pass away after I do. Then I think: My parents each died. Our last minister, Stewart Olson, the one before Bill Babcock, he died six years ago. I guess if they all did it, so can I." She paused. "As Ecclesiastes says, everything has its season—a time to be born and a time to die.

"And as a Christian, I believe in a hereafter, that we return to God. What form that takes no one can really say." Barbara grinned. "It's not like I'm expecting to get on the up escalator and be delivered to paradise. Or find angels there playing harps. I was never one for airy music."

"I want to believe in an afterlife, but sometimes it's hard to imagine," I said.

Barbara's tone turned grave. "Of course, I also have doubts. Everyone who believes has doubts if they're honest with themselves. I suppose it could all be an illusion, that God is an invention of the human mind, and after our life on earth, there is just nothing. But deep inside, it doesn't feel that way at all."

For a while we shared silence. Then Barbara shifted into her conspiratorial tone.

"I've been doing a lot of planning—and some plotting. You know I was Bill Babcock's girl Friday, and now he needs a new permanent one but seems unable to choose. I told him I'm dying, so he has to. Very subtly, I'm steering him to Liz Cook. Liz is efficient, organized without being bossy, and, though younger than Bill, not likely to make mischief in his marriage. That's my plan

for the church." Her tone changed again. "Then I'm trying to figure out whether I should change the will I wrote last year."

"How so?" I asked.

Barbara explained that her sister's daughter Anne, the niece
from Rhode Island, was also Barbara's goddaughter, and she felt
responsible for the young woman.

"When my sister died, Anne lost her moorings." Barbara
stopped abruptly. "Do you really have time to hear this?"

I looked at the clock on the far wall. It was getting late. I had
been at work for ten hours. Mondays were always the longest
days, when I faced what had happened over the weekend and
launched the new week. But I thought to myself: If I were in the
middle of a procedure, or administering a treatment, the late
hour would be irrelevant. I was, in Barbara's words, providing
the medicine of friendship, and it felt as important at that moment as anything I could do or give. And, I admitted to myself, I
was curious. I nodded.

"Anyway, I'll cut to the chase. Anne was never an easy child.
Her parents divorced, and my sister did the best she could. After
her mother's death, Anne fell in with a fast crowd. I tried to discipline her. But she was nineteen, away at college. She never finished school, ran off and married an older man no one in the
family approved of."

I wasn't sure what to say, so I offered platitudes. "These are
difficult problems to solve. I'm sure you tried your best and had
her interests at heart."

Barbara's gaze bored into mine. "That's what I told myself at the time. But now I have to confess that's not entirely true."

I shifted uncomfortably in my seat. Barbara and I had discussed our families before, how hard it was for her to lose her husband, how my father's unexpected death was devastating for me. She had also asked about Pam and about our children, and she was particularly intrigued by how my wife and I struggled to balance work and home life. But the conversation had never ventured into current conflicts.

She went on, "Yes, the man she went off with proved to be every bit the unsavory character I suspected. Lazy. Couldn't hold a job. A drinker. So I was right, and Anne was wrong. But what good did it do to be right? Over the years Anne became more and more distant from me and her cousins, until we hardly spoke. A card at Christmas." Barbara's voice grew heavy. "I didn't make it easy for her to stay close to us. It was my pride, stubborn pride."

Barbara said that when her niece had appeared in the hospital, neither had much to say. Then they talked about the baby. Barbara said she'd also probed, raising money issues with Anne, and she was convinced her niece wasn't visiting as a gold digger. "Finally, we talked about what happened. What Anne did, what I did, how I didn't make things better. She came to tell me she's decided on a divorce. Funny, I don't feel much satisfaction in being right.

"Now it's my job to try to reconcile her to the rest of the family."

That week Barbara left the hospital for home. Hospice nurses attended to her, and I was in regular contact with them. For a while Barbara was able to talk on the phone. She updated me on her "plotting." Liz Cook was chosen to replace her at the church. She was making progress in reconciling her niece with her cousins. Her lawyer had redrafted her will.

It was not long before the cancer overtook her. Reverend Babcock was at Barbara's side when she died. He called me and said that she'd had several hours of labored breathing but had been deeply sedated. The nurses were wonderful, helping her pass without apparent pain.

I thanked him for calling, hung up the phone, and sat in silence. Although I had expected this outcome for quite some time, I felt a gnawing pain of loss. I accepted that medicine had its limits. It was just that I cared for her so much; it was impossible not to. But I also felt deep gratitude. Barbara had opened herself to me in a way no patient had before.

A patient's revelation of her deepest feelings and thoughts is one of the most precious gifts a doctor can receive. It has happened with me when I have reached the level of relationship I did with Barbara, of friendship beyond the professional.

About three decades ago, Dr. Elisabeth Kübler-Ross probed patients' intimate thoughts and feelings in the context of research rather than friendship. Kübler-Ross, a psychiatrist at Billings Hospital in Chicago, was teaching divinity students about caring for the terminally ill when she realized that there were scant data upon which a structured approach could be built. After some resistance, Kübler-Ross received permission from the medical staff to interview very sick and dying patients.

Kübler-Ross conducted individual interviews in a special room, while the student audience, which could number fifty or more, observed through a screen-window. The interviews were open-ended, and most patients welcomed them eagerly. At the time, death was largely a taboo subject, and the burden of silence weighed heavily on those facing their end. Kübler-Ross's book based on these interviews, *On Death and Dying: What the Dying Have to Teach Doctors, Nurses, Clergy, and Their Own Family,* was groundbreaking. It was the first detailed account of the emotions of the dying. As a result of her work, the severe isolation of the very sick became a subject as urgent as their physical suffering.

Kübler-Ross delineated five stages through which a dying patient progresses. First there is denial, from the shock of the diagnosis—the doctors were wrong, the X ray was misread, the biopsy was misinterpreted. On the heels of denial comes anger, at the unfairness of the way the world works and, for those who are religious, at God: "How could he do this to me?" Desperate to escape their apparent fate, patients enter into what Kübler-Ross termed "bargaining." The secular bargain with their doctors and families, while the religious bargain with God, in each case offering something in exchange for a reprieve. Usually, it is a promise—for the religious, to attend church or to give to charity; for the secular, to submit to any test or procedure requested by their physician. When no reprieve appears, the next stage sets in: depression. The patient's energy wanes as he feels the full impact of his prognosis. Under this strain, the patient can no longer avoid reality, and the fifth stage, acceptance, sets in. Yet, Kübler-Ross contended, acceptance is still marked by the illusory belief that a miracle will occur, that at the very last minute a cure will arrive and death will be avoided. This belief she called "hope":

> [T]he hope that occasionally sneaks in [is] that all this is just like a nightmare and not true; that they will wake up one morning to be told that the doctors are ready to try out a new drug which seems promising, that they will use it on him and that he may be the chosen, special patient. . . . It gives the terminally ill a sense of a special mission in life which helps maintain their spirits. . . .
>
> No matter what we call it, we found that all our patients maintained a little bit of it and were nourished by it in especially difficult times. They showed the greatest confidence in the doctors who allowed for such hope—realistic or not—and appreciated it when hope is offered in spite of bad news.

I deeply respect what Dr. Kübler-Ross accomplished in breaking the taboo around the discussion of death. But many whose work is primarily with the dying have taken issue with her paradigm. In my experience, there is not always such a discrete unfolding of five stages leading to acceptance. Sometimes denial

persists to the end. Other times anger is unrelenting. And I dis-
agree most of all with her definition of hope. I do not hold out
the prospect of an illusory last-minute reprieve, as Dr. Kübler-
Ross suggests. Frances Walker's case had taught me otherwise.

Kübler-Ross is correct that we should not give up on any
patient, terminal or otherwise. Early in the illness, the right to
hope against all odds, to seek a miracle, is valid, as in George
Griffin's case. There is a time when the inevitable must be
accepted, but this step need not extinguish hope.

Barbara Wilson's unique calm and acceptance were present
from the outset—not surrender but steady realism; she set the
parameters on her care with a clear-eyed vision of what was possible, what made sense to her, how she wanted to live, when it
was time to die. She seemed to be always in control, of herself
and her circumstances, dictating her own terms. And she never
relinquished a vision of the future, even when she knew she
would be gone. Barbara did not cling to a desperate belief that
I would arrive at the bedside with a cure from the laboratory
just in the nick of time, though I deeply wished such a moment
had occurred—for her and for the innumerable others I have
cared for. The kind of hope she showed me was very different
from what Kübler-Ross described. Barbara's hope was real and
undying.

In her case, it reflected the fact that she had found purpose and
created meaning in her life through relationships with her loved
ones, and with her God. Barbara did not dwell on the ineffable

questions "Why me?" and "Why now?" She saw death as a natural part of life. It is not necessary to be a person of profound faith to hold this view and to act on it. I sensed that my father had the same approach to life and to death. Personal philosophy and experience, rather than faith, seemingly prepared him to acknowledge the reality of his mortality.

When I was entering my teenage years, on weekends my father and I would take long slow walks in the early evening around our neighborhood in Queens. We often spoke about school, or world events, or the family. But on occasion my father would talk about death—specifically, his death. These were frightening moments for me, but I knew he did it for a reason. In the background was the loss he had suffered in his twenties, when his father died abruptly of a stroke, and the random carnage he saw as a soldier in France during World War II. He was far from solid in his faith and deeply doubting about an afterlife. But like Barbara, he understood death as a natural and inescapable part of life. And like Barbara, he was driven by hope despite this unalloyed acceptance of mortality. Uncertain of God, he looked to love and how it would shape the future of his family. Uncertain of an afterlife, he believed in the persistence of memory to make his presence palpable when he was gone. Was he preparing himself by preparing me?

My work regularly brings me in close proximity to death. Like every doctor, I have learned how to compartmentalize the fear and anxiety it naturally provokes, in order to function effectively at the bedside. Yet there are times when thoughts of my own mortality break through, and I ponder how I will face it. I, of course, will not know until the time comes. I imagine conjuring the words and actions of Barbara Wilson, and retrieving the voice of my father from those evening walks. The idea that I might feel hope at the end gives me a sense of deep comfort.

CHAPTER 6

Exiting a Labyrinth of Pain

In the autumn of 1979, while training for the Boston Marathon, I ruptured a lumbar disc. When the pain did not immediately subside, I underwent a discectomy, a limited operation that removed the edge of the bulging disc. The surgery did not fully return me to my prior state. When I tried to run, a dull ache grew in my back and hip.

Six months after the surgery, I moved from Boston to Los Angeles. One Sunday morning after breakfast, I stood up from a chair and collapsed. My lower back felt as if it were being twisted in a vise, and electric shocks sped down my buttocks into my legs.

X rays showed no clear cause for the pain. There were no ruptured discs. I saw many consultants: rheumatologists, neurosurgeons, doctors of sports medicine. Each told me that the

THE ANATOMY OF HOPE

lumbar spine is a "black box" and often a mystery in terms of persistent pain. I kept seeking clinical opinions. Finally, an orthopedist in Beverly Hills told me what I wanted to hear: "Your spine is unstable," he said. "I can fix it with a fusion. You'll be running in a few weeks." A fusion, done by harvesting bone from my pelvis and inserting it like a bridge along the sides of my lower vertebrae, would create an internal brace and remedy my lumbar instability. The heady promise of running again erased any other consideration.

I awoke from the operation in the intensive-care unit, red blood cells being infused into my arms. As the anesthesia lightened, I was hit with excruciating pain throughout my lower body like none I had ever experienced.

Even more terrifying, I could not move my legs. They seemed to be frozen. No one was sure why. The orthopedist postulated that the blood spilled during surgery had inflamed my nerve roots, causing the severe pain and preventing my legs from functioning normally. I was taken home in a body brace, a large metal cagelike structure that fit tightly around my lower torso. It prevented me from moving too far in any direction and stressing the bone grafts. But it was hardly necessary to artificially immobilize me. The pain was so severe that I could not stand or sit. I was in bed all day and all night, sleeping a few scant hours only with large doses of painkillers. Yet even narcotics could not stop the relentless pain in my back, buttocks, and legs. For months I lay on ice packs, the numbing cold affording a few moments of relief.

When I returned to see the orthopedic surgeon, I asked why I wasn't getting better. His words wove a tight narrative of what I could look forward to: "You must have developed scars around the nerves from all the inflammation. With the nerves tethered, every time you move quickly, or twist or turn, they get tugged, setting off the pain."

My throat closing, I asked what could be done.

"I can operate again and try to cut away some of the scar."

Wouldn't that cause further inflammation and new scars?

Probably yes.

I returned home feeling as if I had entered a maze with no exit in which I would wander from one point of pain to the next.

I began physical therapy at a local facility. Most of the people in the treatment area were elderly, with severely arthritic hips and knees that had been replaced. The staff was compassionate and encouraging, but the therapy was largely passive. I would lie on my back, and the therapists would slowly move my legs up and down. Then they put me in a warm pool with submerged parallel bars. I took step after faltering step, learning to walk again.

A stretcher was moved into my laboratory, and I spent most of the day flat, reading scientific articles and planning experiments with my technicians. I learned how to do basic things in ways that didn't set off pain—how to get up from a chair, turn my torso, reach down to pick up something from the floor. After nearly two years, I could swim four or five laps and sit on a stationary bicycle to pedal some minutes.

In the quiet of the day, the orthopedic surgeon's words echoed in my mind, that the scar would encase my nerves for life. Disgusted with myself for undergoing the fusion surgery, I decided to stay away from doctors. One had badly hurt me, and none could help me. I would have to live within narrow limits.

And so, for nineteen years, I did. The boundary on my life was like an electrified fence at the perimeter of a prison: If I ventured too far and tested it, I was thrown back from the shock of its force into my confinement. If I stayed inside the fence, I was somewhat safe. So, I assessed each movement and task of the waking day for its distance from that perilous perimeter. When I awoke in the morning, I would shift my legs over the side of the bed and cautiously lift my trunk, using the force of my forearms to avoid a pull on my back. In my laboratory, I would lean at a minimal angle to peer into the ocular of a microscope, or to inspect cells in a petri dish, because quick flexion was risky. On ward rounds, I would sit down every ten minutes or so to rest my back.

When my sons were in grade school, I wanted to teach them to play baseball, but swinging a bat triggered sharp pains in my back, and I retreated to the sidelines. When my four-year-old daughter climbed monkey bars and jungle gyms, I worried that I lacked the resilience to catch her if she fell. I had been an avid walker, strolling the neighborhood in the evenings for an hour or more with Pam, but now, after I walked four or five blocks, my hips and back would cramp, forcing me to abandon the excursion.

There were periods of respite, months of relative comfort, but then an apparently innocuous movement would cause my lumbar area to explode in spasm. I never knew when or where it might happen. Along with unpredictable pain came its companion, a sense of prevailing fear.

In the summer of 1999, such an unpredictable event occurred. A close friend, aghast at my restricted mobility, had encouraged me to try back massage. I was leery of allowing anyone to touch my spine, and my concern was borne out. Despite the licensed massage therapist's good intentions, he was too aggressive in his work. I left the session with a growing ache, and by the end of the day I was gripped by vicious pain from the base of my neck to the bottom of my coccyx. Days passed, and the muscle spasm did not relent.

Frustrated, in deepening pain, I sought advice from a rheumatologist, an expert in bone and joint disease, at the Beth Israel Deaconess Medical Center where I work. He examined me and said that the muscles in my back had been traumatized by the massage; with rest, anti-inflammatory medication, and tincture of time, the acute pain should quiet down, and I would return to my previous level of functioning. But he made me promise that when this flare-up abated, I would see a physician at the New England Baptist Hospital who specialized in rehabilitation medicine, Dr. James Rainville. "He may be able to help you in a more fundamental way," the rheumatologist said. "The Baptist was where Larry Bird was put on his feet after his back problem." I nodded politely but did not believe that anything substantial could be done.

Like all Celtics fans, I had closely followed Bird's condition, especially given my years of debility. He had undergone spine surgery and returned from rehabilitation still a star. And basketball was an unforgiving sport: The impact, as you fought for rebounds under the boards, took jump shots from the three-point line, and made fast breaks, would tax even a normal spine. Bird seemed more than able to keep up.

Contrary to the rheumatologist's prognosis, my back spasms did not fully subside. In fact, I had episodic pains not only in my lower back but between my shoulders and in my neck. I moved through the day with even more tension and worry than before, and I shied away from any activity that might exacerbate the muscle spasms. I went for an MRI scan of my spine, but it did not reveal a cause for the muscle spasm that gripped me from my neck to my hips.

Weeks passed, and my despair deepened. I came home from work exhausted by the pain. For almost two decades, I had tried to achieve an uneasy truce with my condition. Now it seemed as if I had violated the terms of the treaty, disregarding the limits that were set for me. Because it didn't make sense that the pain would be lasting for weeks, it became even more frightening. Pain can be a volatile and fickle force and, when provoked, might unleash a torrent that overwhelms the initial provocation. I knew of cases in which people suffered with no obvious explanation.

All of my former strategies and remedies, resting in bed on ice packs, taking Valium for muscle spasms, gently stretching my legs and torso, were to no avail. With the sense that I had nothing to lose but also that my plight would only be confirmed, I visited Dr. James Rainville at the Spine Center of the New England Baptist Hospital.

The New England Baptist Hospital is a stately old brick building at the crest of Boston's Mission Hill. Its large lobby is appointed much like a traditional Yankee living room, with muted carpeting, upholstered wing chairs, and an antique grandfather clock. A long, narrow corridor leads from the lobby to the clinics and operating rooms. Along this passageway, the history and achievements of the hospital are presented in a series of enlarged photographs arranged chronologically and annotated like a display at a museum.

As I walked down the corridor, my eyes stopped at a picture of a young Senator Edward Kennedy. He was lying in bed, his head propped up, offering a warm smile of relief. The text adjoining the picture explained that, as a young man, Kennedy had been in a plane crash and had fractured his spine. He had come to the Baptist Hospital for rehabilitation. I tried to recall recent images of the senator that might indicate continued debility. But newspaper pictures and TV footage showed Kennedy standing tall in Congress, deftly navigating his boat off Hyannis, and walking briskly on the campaign trail. There was no hint that he was once bedridden with a fractured spine.

A few short steps down the corridor was a photographic tableau of Larry Bird. In the first photo, he was seated next to a smiling doctor in surgical garb. An adjoining picture showed Bird in the midst of physical therapy. The text explained that the celebrated Celtic had been successfully restored to the parquet of the Boston Garden. I lingered, studying Bird, his physician, and his physical therapist, then proceeded to the Spine Center on the fifth floor.

As I changed into a hospital gown, I cursed myself once again for my stupidity in letting anyone touch my fragile back. My thoughts were broken as Dr. Rainville strode into the examining room. He was a tall, athletic man with thick, graying eyebrows and the rugged good looks of a Hollywood gunslinger. He greeted me with a firm handshake. "Tell me the story from the beginning," he said.

I recounted the past nineteen years—the running injury, the fusion surgery, the many precautions of my days, ending with the massage and the unrelenting spasms.

Dr. Rainville listened intently, his face registering no reaction. He began a detailed physical examination, testing each muscle and joint individually for strength by pressing with his palm. Then he applied a calibrated instrument that resembled a protractor, and took exact measurements of the range of rotation of my hips, spine, and neck as I moved forward, bent backward, and turned to each side. My muscles tensed as I performed each of these maneuvers, and my jaw tightened in fear that the physical examination itself might release a shower of pain.

Dr. Rainville planted the MRI scan of my spine on a lighted box on the wall and systematically inspected the film, vertebra

EXITING A LABYRINTH OF PAIN

by vertebra. Seemingly satisfied that he had reviewed all the relevant data, he turned and stood before me.

"You are worshiping the volcano god of pain," he declared. "The volcano god of pain is your master."

I had been warned that Rainville was a brash, in-your-face clinician who held contrarian views. But what on earth was he talking about?

"What do I mean that you are worshiping the volcano god of pain?" he asked. "You interpret pain as a red flag, a warning that you are doing damage to your body. So you sacrifice things that you love, activities that give your life joy, to be kept free from pain. You say to the volcano god: 'I will give up walking long distances if you keep me out of pain. I will give up lifting my children if you keep me out of pain. I will give up travel, because long trips stress my spine. Just keep me from pain.'

"But this god is never fully satisfied with any offering: It is appeased for only a short while. So the more you sacrifice, the more the god demands, until your life contracts, as it has, into a very, very narrow space.

"I believe you can be freed from your pain. I believe you can rebuild yourself and do much, much more."

I studied James Rainville's face. He had the burning eyes of a zealot, and he read my thoughts.

"Bullshit talks, results walk," he said. "You think what I am saying is complete bullshit. You've lived all these years without any real hope, and it's hard to open that door and glimpse a different kind of life." He paused and then spoke with gravity. "It's your choice: to try or not to try. You can walk out of my office now and believe everything you've believed for the past nineteen years, and live the way you have. Or you can test me. And I'll tell you now, I'm right."

I remained mute, unsure how to respond to this barrage. My thoughts raced: He doesn't understand my condition, the immutable scars from the surgery. He is dangerous, threatening to upset even the little equilibrium I've established over the years. But then a strange, powerful thought welled up: *Could he be right?*

"How can I test you?" I finally asked.

"Ignore the pain," Dr. Rainville shot back. "No more worshiping this god. No more sacrifices. Just disregard its demands. The pain doesn't mean anything serious. As your mind reorients its beliefs, the pain will lessen. Once that happens, you will be able to rebuild your body. It won't be easy or quick or without setbacks. The muscles in your low back are at thirty percent of their normal tone. Your ligaments and tendons are contracted from lack of use. They carry a memory of the trauma done to you, and they recall the pain at the slightest provocation. By using your spine, muscles, and ligaments, and strengthening them, you erase that memory. You can compensate for the damage that was done and live a normal life again."

Rainville explained the basics of the program he was prescribing. My muscles would be challenged steadily with increasing weights until they were "reeducated" to relinquish their memory of past pain. Despite what I would feel during this process— almost certainly an increase in the muscle spasm, and release of the familiar electric shocks over my buttocks and thighs—I was not at risk of being hurt. Everything that was being done to my body would serve to rewire my brain until strength and endurance were restored and the perception of pain erased.

My heart was racing as I exited the hospital. It was strange: There was a perverse comfort in inertia. As a physician, I had learned the need for hope. Yet I had completely abandoned it. Despite education and knowledge and experience, when you are the patient—suffering, confused, and despairing—it is very, very hard to take matters into your own hands. I was not a George Griffin, able to stand alone and challenge the prevailing assumptions. I needed an external voice, strong and determined, to guide me.

But what if Rainville was wrong? His arduous rehabilitation program might end for me with worse pain. A mere massage had triggered weeks and weeks of suffering. What would even greater stresses do?

And what was he *really* saying, in cold biological terms, not flamboyant metaphors of volcano gods? That changing my beliefs could dampen flaming circuits of pain? Was I some mystic or yogi who could walk barefoot over hot coals or lie down on a bed of nails?

As the days passed, I kept returning to Dr. Rainville's words. I pictured resuming activities so long gone. I imagined myself stronger, able to walk comfortably, to lift, to travel, to play sports. The power of the possibility became too much to resist.

With trepidation, I enrolled in the Baptist's rehabilitation program. I was alarmed when the physical therapists took me into the treatment room. It was a large, open space filled with what looked like instruments of torture. A group of patients was in the midst of therapy, doing things that seemed designed to cause, not erase, pain: picking up milk crates filled with lead bricks; brusquely rotating their torso while holding a heavy medicine ball; performing a back press against some hundred pounds. Was Rainville out of his mind? How I could ever risk such maneuvers? The thought of it terrified me.

But deep down flickered the desire to fight.

The therapist prescribed three weeks of daily stretching at home before I could take part in formal rehabilitation exercises. If I acted too quickly, my muscles would be overwhelmed, and I might abandon the attempt. He instructed me to begin slowly, to move bones and joints that had not been tasked for years. This in itself proved exhausting; my body was so feeble that even lifting my legs up against gravity broke a sweat.

It was the beginning of the school year when I went back to the treatment room. Boston in early September pulses with the steady pace of young men and women returning to college, nervous anticipation on their faces. I told myself I was a new student, ready to reeducate my mind and body. But it didn't prove so easy. Even the first lessons in Rainville's classroom were daunting. Each exercise caused shooting pains in my back that ran down my buttocks, thighs, and feet. The therapists were unmoved by my pains, telling me firmly to keep challenging my body. I would leave the sessions depleted, needing to rest for hours, lying in bed on ice packs to numb the protesting nerves and muscles.

When the pain and exhaustion from the maneuvers were at their worst, when doubt raised its strident voice and said Rainville's talk *was* bullshit, I summoned a dreamy vision of the future: walking hand in hand with my daughter to a pond some two miles from our home to feed ducks and search for frogs; emerging from an airplane feeling strong and ready to explore a new city; dancing a traditional Eastern European circle dance at a family wedding.

Each fantasy released a current of warm, soft energy that suffused my body. My exhausted muscles and taut ligaments seemed deeply nourished in its wake. My spine, in its cacophony of pain, became distinctly quieter. These physical perceptions made me feel foolish and perplexed but also curiously intrigued.

Over three months or so, the constant pain became intermittent, then rare. "The body has a remarkable ability to heal," the physical therapist had said to me at the start of one session. Those words conjured a very different picture of what existed in my body. Yes, as the orthopedist said, there had been inflammation and probably some scarring. But it was not irremediable. There were ways for the muscles, ligaments, and tendons to become stronger and more supple at the same time, to insulate the nerves from the strain of sharp movements. Even when the nerves spoke up, the messages they sent to the brain were modulated, their anger tempered, so that severe pain did not result. Over time these messages, once bellowing, could become a whisper, and then fall largely silent.

I realized that the surgeon had said to me what I had so ineptly said at times in my career to others: that there was no hope. Medicine is too uncertain a science, biology too variable, to have that kind of hubris.

I began to walk—five blocks, then ten, then fifteen, then a mile. I challenged myself to travel longer distances, to hike one or two steep hills. I felt more at ease with and more eager about my rehabilitation exercises. Advancing to each new level caused days of spasm and pain, but I tried to ignore it all, repeating Rainville's mantra, that my body was relinquishing an old form of memory and acquiring a new narrative.

I did discover that there were limits, that I was vulnerable to injury. I joined an aerobics class called "Funky Groove" that involved doing 250 jumping jacks while earsplitting techno music blared in the background. The pounding proved too much for my muscles and joints and caused persistent pain that did not resolve with the Rainville regimen. So I quit the class. I realized I was not twenty-seven years old anymore. But I also told myself that I could live happily within my new boundaries. I had gained, above all, the sense that I had some level of control over my choices.

After a little over a year, the daily pains all but passed. I awoke in the morning unafraid, moved through the day comfortably, and fell asleep with a prayer of thanksgiving on my lips. I felt reborn. It seemed almost magical.

CHAPTER 7

The Biology of Hope

After my recovery, I thought about what I had learned from my experience. I tried to see how it could give me a more profound understanding of my patients and how to help them. There were obvious differences between my situation and theirs. I was not afflicted with a potentially terminal disease and did not have to contemplate the possibility of impending death. Nonetheless, there were important similarities. My patients often must pass through painful and debilitating therapies for any chance of achieving a remission. And, as with Richard Keyes and Dan Conrad, the hardest part can be taking the very first step. Fear is often the most significant hurdle—fear of pointless pain and suffering. For my patients, as for me, it was hope that inspired the courage to overcome fear, and solidified resilience during an arduous treatment.

I was most intrigued by the sense that I may have felt physical changes caused by hope. But I distrusted my impression. Scientists view personal anecdotes and impressions warily, and rightly so.

So I asked, as a scientist, is there a biological mechanism whereby the feeling of hope can contribute to clinical recovery? And if there is such a biology of hope, how far is its reach, and what may be its limits? Or is hope a feeling that occurs along with certain physiological changes but has no causal link to them?

I set out on a journey to answer these questions, for my patients and for myself. Frankly, I was skeptical that I would learn anything substantial. Much of what I had heard about mind-body connections was the stuff of self-styled gurus, long on claims and short on rigorous scientific data. These gurus portrayed the mind as dwarfing the body, so that with enough desire and determination, any disease or debility could be overcome. Their subliminal message was seductive: If you only knew how to will it, you would be in complete control over your condition.

This stance was brought into sharpest focus in popular books I had read featuring unexpected remissions of dreaded diseases. Instead of portraying the remissions as mysteries that merited serious investigation, sweeping conclusions were drawn—that lupus had given way to laughter, that cancer had yielded to meditation. How could the tellers of these tales so surely assert this? How many people had laughed and still suffered from lupus? How many had meditated and still succumbed to cancer?

During thirty years in medicine, I had seen many patients trying to shrink their metastases through these modalities, yet the
tumors grew. For these cases, the proponents of the mind-body
connection had an explanation that, to my mind, worsened the
plight of the patient. The reason that visualization and meditation
failed to eradicate breast cancer or lymphoma or a brain tumor
was that the afflicted individual was not doing it well enough; he
was holding on to the anger and depressive thoughts that had
activated the malignancy. These negative emotions blocked the
salutary effects of positive thinking. In essence, invoking a mind-
body connection in this way blamed the victim for his own malady and ultimately for his demise.

As unconvincing as such arguments seemed, I was loath to
discount the possibility of the mind altering the body. The mind
is a product of the brain, and the brain and the body are inti-
mately connected by a network of nerves and by hormones and
other active molecules that mediate many physiological functions. The nexus of mind and body was glossed over in my
education. I had been taught a traditional medical curriculum
in which each organ—liver, kidneys, heart, brain—and each
system—circulatory, endocrine, respiratory, neurological—was
approached in an isolated, reductionist way. Lip service was
always paid to a larger whole, but the physiology of the body
was linked with the workings of the mind only in rare instances.
Stress was agreed to be a bad thing. There were a few references
in my courses to the deleterious effects of anxiety and despair:
The type-A personality, who was relentlessly hard-driving and

never relaxed, was predisposed to heart attacks and ulcers and gastritis; clinical depression was viewed as among the causes of disrupted menstruation and infertility. But beyond these few examples, the mind was kept segregated and not seriously considered as an influence on clinical outcomes. Even years later, when I was an established doctor, there was scant serious study of the effects of positive emotions, like hope. Yet there *had* to be a biology of hope, as there was a biology of fear and anger and anxiety and every other emotion. The fundamental question was not whether such a biology existed, but what did it encompass?

My quest began the year after I completed my rehabilitation program. It became a passion, an intense, obsessive drive yielding the same excitement that I feel when I begin a new project in my laboratory or test a new therapy in the clinic.

I embarked with both the advantages and the disadvantages of a neophyte. I began speaking with experimental psychologists and psychiatrists, neurologists, experts in brain imaging and chemistry, specialists in mechanisms of pain, in metabolism and muscle physiology.

Some of my first discussions were with Dr. Bruce Cohen, a professor at Harvard Medical School and the director of McLean Hospital, one of the preeminent institutions for psychiatric care and research in the United States. A sharp right turn off Main Street in Belmont, Massachusetts, brings you to McLean's sprawling, tree-lined campus. There are small brick buildings and clapboard cottages dating from the nineteenth century, and a few modern concrete structures, giving an aggregate impression of a small New England college.

Dr. Cohen's office is housed within the main administration building. It is a large, open space with tall windows, an Oriental rug on a hardwood floor, a well-kept fireplace, a yellow uphol-stered couch, and wing chairs. Over the fireplace is an oil portrait of the eponymous benefactor, Benjamin McLean, a Boston grandee with shocks of white hair and a sharp nose like the prow of a ship. In the early 1800s, after the founding of the Massachusetts General Hospital some ten miles away in downtown Boston, McLean bequeathed the money to support the psychiatric hospital that was then given his name.

Bruce Cohen, a short, cherubic man with thinning gray hair and bright blue eyes, offered me a seat in one of the wing chairs. Instead of retreating behind his vast antique desk, he settled himself into the corner of the couch. We exchanged pleasantries, and I commented on the portrait of McLean over the fireplace.

"He was a visionary," Cohen said. "He believed that the work-ings of the mind, in health and in disease, would in essence be no different than the workings of the body."

I had prepared myself for the meeting by reading some of Cohen's research papers. It is relatively easy these days to access the lifework of any scientist. A few clicks of the mouse link you to search engines of the National Library of Medicine, the com-pendium of all published biomedical studies. Dr. Cohen's focus was on the biological basis of drug dependency.

"So how can I help you?" he asked, an open-ended question that classically begins the doctor's interview of a new patient. I set out my objectives: to learn as much as I could about how a complex emotion like hope was formed in the mind, how it might affect the chemistry of the brain, and how it could change the physiology of the body. I briefly told the stories of patients like George Griffin and Dan Conrad, and the story of my recovery. Dr. Cohen's expansive face narrowed in intense concentration; he bobbed his head in rhythm with my speech, as if absorbing each word for its meaning.

He began, "Nothing you describe about how hope may affect pain and debility, as well as risk taking and resilience, is outside the realm of science. In fact, our research team here is charting circuits in the brain that are surely involved in part of what you and your patients experienced. All of us in the field are just scratching the surface, and one day we'll look back and realize how little we really understood. But that's how science always begins. My own aim, to understand how dependency and addiction occur, touches on which chemical transmitters and neural circuits in the brain are important in motivation, in setting goals and seeking rewards, and in sustaining behaviors—all relevant to your question."

Dr. Cohen leaned forward in the posture of a teacher instructing a student. "For purposes of further discussion, we should define the terms 'mind,' 'body,' and 'soul.' Our emotions, thoughts, perceptions, reflections, and desires, all activities of the mind, seem to exist as if they were disembodied, hovering in an ether several inches above our heads. Of course, this is not the case. 'Mind' is a manifestation of brain. What we view as products of the mind—thoughts, feelings, and emotions—are a mighty mix of chemicals and electrical circuits that have evolved over the millennia, and still are changing. So, too, is consciousness—our memory of the past, awareness of the present, and anticipation of the future. 'Body' includes brain, and thus mind, so that the construct 'mind-body connection' only emphasizes the artificiality of how we have traditionally divided them."

I was surprised, I said, that he had included "soul" in the troika of terms to define.

"Soul is fundamentally a metaphysical and religious concept, where the divine spark resides. For the agnostic, the soul is in doubt; for the atheist, it is a fiction. But for a person of faith, the soul has permanence, while the body—including brain and mind—does not. The soul is not a product of cerebral chemicals and neuronal circuits." As scientists, he went on, we acknowledge that there is no way to experimentally locate or characterize such a metaphysical entity. It was beyond science; it was a matter of faith. But, Dr. Cohen added, as a physician who believed in the soul, he sought to address it in his patients. I thought to myself how pleased Barbara Wilson would be to hear a doctor like Bruce Cohen give the soul its due.

Dr. Cohen paused thoughtfully and said that I should begin my exploration of the biology of hope by learning about the placebo effect. Belief and expectation, two cardinal components of hope, were also fundamental to the biological effects of placebos.

I was taken aback: "Placebo" in my mind denoted something false.

"You'll be surprised to learn the true effects of placebos. Analyze the experiences of your patients and your own experience in enduring pain and debility, as if you were subjects in an experiment on placebos. You'll see how informative those experiments are to building a model of a biology of hope."

The term "placebo," I learned from articles on the subject, is Latin for "I shall please" and is derived from the Catholic vesper service for the dead. In the past, paid mourners would participate in the service to pacify the grieving family.

In its initial meaning, a placebo, like a sugar pill or a saline injection, was given by a doctor to appease demanding or desperate patients. A physician who prescribed placebos was, in essence, a charlatan, the medical equivalent of P. T. Barnum. The use of placebos in medicine changed in the 1950s with the introduction of scientific clinical trials. Researchers testing experimental drugs would administer either a placebo or actual therapy to two groups of patients. The patients receiving the placebo served as controls. The inert placebo was meant to mute the background noise of body processes, the random and supposedly meaningless oscillations that occur in our physiology. The outcome among the placebo-treated control patients was judged as the zero baseline and subtracted from the observed effects of the true therapy.

Recently, though, a number of scientists have questioned the assumption that placebos were inactive, contending that they have significant biological effects and, as Dr. Cohen indicated, provide one of the clearest windows into the nexus between mind and body, particularly on how belief and expectation may affect pain and physical debility.

PLACEBOS, PAIN, AND DEBILITY:
THE EXPERIMENTS

To understand the experiments on placebos and pain, we need to know about the genesis of pain. Imagine a stubbed toe or a knife cut on a finger. Specific nerves in the tissues carry signals of pain to a part of the spinal cord called the "dorsal horn." The pain signal leaves the dorsal horn, the receiving area, and ascends the spinal cord at several points, as one runner in a relay race passes the baton on to the next. The finish line is in the brain, where we consciously perceive the pain.

Within the spinal and brain cord are cells that can turn pain signals on or off. They are aptly termed "on" and "off" cells. On cells increase pain and appear critical in withdrawal reflexes: The output of the on cells causes us to pull our hand off a hot burner or jerk away from the edge of a sharp knife, even before we consciously sense the pain. The off cells act like circuit breakers: They interrupt the flow of painful signals. If off cells were dominant, then we would not be primed for pain; rather, we would move numbly through the world, as if encased in armor. This would be a dangerous state, given that pain protects us from

potentially damaging activities. So the on cells normally domi-
nate and restrain the off cells, and we are wired to perceive
pain—and to pull our hand off the burner or away from the
knife.

Certain powerful drugs derived from the opium poppy, like
morphine, can block pain. They work by throwing up roadblocks
in the central nervous system and, in effect, stopping the relay
runner from passing on the baton. An injection of morphine
works by shutting down the on cells. When on cells are shut
down by opiates, the off cells are free to block the flow of painful
signals. For a while, our neural circuits are switched off, and we
function as if we were indeed wearing a suit of armor.

What does this have to do with hope? It turns out that we have
our own natural forms of morphine—within our brains are
chemicals akin to opiates. These chemicals are called "endorphins" and "enkephalins." Belief and expectation, cardinal components of hope, can block pain by releasing the brain's endorphins
and enkephalins, thereby mimicking the effects of morphine.
This conclusion has been substantiated by several research
groups. One of the leaders in the field is Dr. Fabrizio Benedetti,
of the Department of Neuroscience at the University of Turin in
Italy.

A classical Benedetti experiment involves a research assistant
causing a volunteer pain by compressing a cuff around the arm,
in effect cutting off circulation with viselike pressure. The volunteer, a young man, winces in response. He is attached by recording wires to a series of monitors that document physiological

changes in response to the painful stimulus. With the pressure, there is increased heart rate, sweating, blood pressure, and muscle contraction. It takes a few moments for the tension in the volunteer's face and limbs to pass when the cuff is deflated.

The research assistant wears a spotless starched white coat and tells the volunteer in an authoritative voice that prior to the next stimulus, an injection of morphine will be given to prevent any pain. A syringe from the researcher's pocket appears. The volunteer's eyes follow the assistant as he picks up the vial of morphine and fills the syringe. The drug is injected. A few minutes are allowed to elapse before the cuff is once again inflated. The volunteer does not wince or otherwise react. The recording devices show no changes in heart rate, sweating, blood pressure, or muscle contraction. Pretreated with morphine, the volunteer feels no pain.

The procedure is repeated several times in sequence, the syringe filled, the morphine injected, the painful stimulus given, with the same outcome: The volunteer feels no pain. The wariness that marked the volunteer's eyes when he sat down in the lab and was hooked up to the monitors is gone. Rather, his mien is relaxed.

SLEIGHT OF HAND

Then the researcher again holds up a syringe, fills it with a clear fluid from a glass vial, and authoritatively tells the volunteer that the injection will be administered. But this time the vial contains saline, which is essentially table salt dissolved in water.

The cuff is inflated. But the volunteer shows no signs of distress or discomfort. The recordings on the monitors do not jump: The heart rate is steady, blood pressure normal, skin cool and dry, muscles relaxed. The research assistant asks the volunteer how he feels. He reports no pain. None? the assistant asks. No, none whatsoever.

How can this be, that a placebo like saline can mimic morphine?

Benedetti's hypothesis is that the volunteer's belief and expectation that he is receiving a potent drug like morphine generate signals in his brain that release his endorphins and enkephalins. These chemicals course through the spinal fluid and bathe the pain circuits, shutting off the on cells, restraining the runners that pass on the baton of pain. With his on cells blocked, the volunteer's off cells snap into action. The off cells throw up their barriers along the pain pathways. The messages of pain from the nerve fibers on his skin where the stimulus was given are not transmitted. They are stopped cold when they enter the first way station in the spinal cord. His brain never receives them.

Benedetti proves his hypothesis in another experiment that again uses a sleight of hand. The researcher again gives the volunteer a painful stimulus, with the expected painful response. Again he tells the volunteer that a powerful painkiller will be administered before the next cuff pressure. Morphine is given, and pain is blocked; this sequence is repeated several times. Then the second sleight of hand occurs. With the next cuff pressure, the researcher fills the syringe with neither morphine nor saline but Naloxone, a drug that blocks the receptors for endorphins and enkephalins in the brain. Naloxone is like an obnoxious bus passenger who not only occupies his seat but spreads out on all the seats around him. There is no place for anyone else to sit down. That is, Naloxone sits on the surface receptors of the brain's on cells, so there are no "seats" available for the endorphins or enkephalins. These cerebral chemicals are still released but cannot dampen the pain pathways, since their receptors are occupied by Naloxone.

Although the volunteer expects that he is being injected with morphine, he shows the effects of pain: His heart beats quickly, his palms grow moist, his blood pressure rises. Naloxone prevents the placebo effect. Although endorphins and enkephalins are released by belief and expectation, they fail to block the pain signals because Naloxone interferes with their contacting the nerve cells.

There are other chemicals in the central nervous system that can modulate pain circuits, amplifying pain. Substance P is one of them, and another is cholecystokinin, or CCK. It appears that CCK works in part by blocking endorphins and thus enhancing pain. Some researchers contend that expectation and belief also interfere with CCK release, thereby enhancing the analgesic effects of endorphins and enkephalins.

INSIDE THE MIND OF THE SUBJECT

Scientists studying the experimental scenario above try to enter the mind of the volunteer and categorize his psychological responses into functional parts. Each visual, auditory, and tactile component in the experiment and in the behavior of the researcher is a cue, perceived and processed by the volunteer. The starched white coat, authoritative voice, statements about the drugs, and deployment of the syringe are all cues that affect the mind-set of the volunteer. The two major changes are belief and expectation: The volunteer *believes* and *expects* that he will receive pain from the cuff and comfort from the morphine. He also *desires* to avoid pain. Belief, expectation, and desire activate brain circuits that cause the release of endorphins and enkephalins, and perhaps the inhibition of CCK.

Dr. Ted Kaptchuk, who studies placebos at Harvard Medical School, sees the environmental cues and the behavior of an authority figure as part of the ritual of medicine that dates to ancient shamans. And, he emphasizes, a change in mind-set can alter neurochemistry, both in a laboratory setting and in the clinic. When we are patients, suffering from pain and debility, we look to our doctors and nurses for the words and gestures that reinforce our belief in medicine's power and solidify our expectation that we may benefit from an intervention. Recent research shows just how catalytic those neurochemical changes can be in the course of certain maladies.

arthroscopic lavage, or the sham placebo surgery. Patients in the placebo group were brought to the operating room and had their knee prepped with antiseptic and surrounded by sterile drapes. The surgeon requested all the instruments normally used in the true procedure, and he manipulated the knee as if arthroscopy were being performed. But with the sham procedure, small incisions were made at the skin around the knee without insertion of the arthroscope. Saline was then splashed to create the sounds of the lavage. The patient was kept in the operating room for the same amount of time as patients undergoing the true procedure, and the recipients of the placebo spent the night in the hospital and were cared for by nurses who were unaware of what had transpired. All three groups—lavage, debridement, and placebo—received the same postoperative care, which included walking aids, a graduated exercise program, and analgesics. Patients were followed for two years; what was assessed included their level of pain and changes in function with regard to pace and distance of walking and other activities that utilized the joint.

As expected, the patients who underwent the arthroscopic procedure experienced decreased pain in the knee and an improvement in function. But the placebo group had an equal benefit. On the evening of the article's publication, I watched the television news. The program featured an older African-American gentleman who was playing basketball for the first time in years with his grandson; he was in the placebo group.

What explained this outcome? Belief and expectancy—experienced while being wheeled to the operating room, listening to the surgeon articulate his requests for instruments, and noting the sounds of the supposedly salubrious lavage fluid sloshing through the knee—likely released the powerful endorphins and enkephalins that Benedetti documented in his experiments. Except this experiment was performed not in a laboratory, on normal volunteers, but at a clinical trial on people with a painful and limiting malady. Pain was the hurdle that prevented these patients from working to strengthen their muscles and ligaments. Once that hurdle could be overcome by the power of the mind, then the necessary rehabilitation could proceed. Without hope, nothing could begin; hope offered a real chance to reach a better end. Hope helps us overcome hurdles that we otherwise could not scale, and it moves us forward to a place where healing can occur.

Not long after the publication of the study on arthroscopic knee surgery, data were presented on an experimental intervention for low-back pain. Surgeons were evaluating a new device that was inserted into a ruptured or degenerated lumbar disc; the device generated heat that was supposed to help shrink the bulging disc and thereby relieve pain and muscle spasm. A placebo group underwent a sham surgery in which small incisions were made in the skin overlying the abnormal disc, but the device was not inserted. As with the study on knees, the patient did not know whether he was undergoing a true or sham operation; all of the visual and auditory operating-room cues that occurred with the true surgery were reproduced by the doctors and nurses during the sham operation. The results of the placebo group were striking: Nearly half of the patients undergoing the sham surgery reported a decrease in pain and an improvement in function. Again, instilling hope appeared to have a sufficient effect on pain to make rehabilitation feasible.

THE BODY-MIND AND MIND-BODY CONNECTION: THE RISK OF A VICIOUS CYCLE

With our glimpse into how the brain affects the body, we can widen our scope to how the body affects the brain. William James, an eminent psychologist of the nineteenth century, postulated that neural input about the physical condition of our tissues was a primary modulator of our positive and negative emotions. Although this argument neglects the capacity of the brain to generate emotions independent of such visceral input, it does highlight an important aspect of how hope may be sustained or snuffed out. When our organs are diseased, our tissues atrophied, our systems failing, nerves from these diseased body parts transmit signals, such as pain, that potently amplify our feelings of fear, anxiety, and despair. The stirrings of recovery in our tissues help generate the feeling of hope. At that moment, clinical treatments can have their most profound impact on our psychology. With each increment in improvement, the body sends more signals that inform the brain of a return to health. As with a climber gradually ascending from a deep and threatening crevasse, each upward step makes it easier to see the end and sustain hope.

So, while we are familiar with the mind-body connection, and researchers explore how thoughts and emotions affect our tissues and organs, it will be important to pursue research on what can be termed the body-mind connection: how input from peripheral parts is sent to the central nervous system and alters the chemistry of the brain to shape cognition and feelings.

I believe that Dan Conrad's story serves as an example of this interplay between brain and body. Why bother enduring? he asked. The escalating pain and sense of suffocation from his strangling lymphoma flooded his brain with the kind of visceral input that solidified his hopelessness. This further reinforced the assumption that his outcome would be no different from that of his suffering friend Tom Kane.

This is the vicious cycle. When we feel pain from our physical debility, that pain amplifies our sense of hopelessness; the less hopeful we feel, the fewer endorphins and enkephalins and the more CCK we release. The more pain we experience due to these neurochemicals, the less able we are to feel hope.

To break that cycle is key. It can be broken by the first spark of hope: Hope sets off a chain reaction. Hope tempers pain, and as we sense less pain, that feeling of hope expands, which further reduces pain. As pain subsides, a significant obstacle to enduring a harsh but necessary therapy is removed.

The body-mind connection should be thought about not only in extreme cases like Dan Conrad's, where a patient is literally suffocating, but in instances where symptoms may be routine and not life-threatening. Among these non-life-threatening but significant symptoms is fatigue.

People with a variety of maladies ranging from cancer to neurologic disorders to circulatory diseases complain of unremitting fatigue. For decades I would simply acknowledge the expected when my patients told me how exhausted they were. Now I look for ways to alleviate some of the fatigue, because even small reductions in that symptom can have a major impact on a patient's sense of hope.

This phenomenon became apparent to me recently when an older woman with an indolent form of chronic leukemia began to despair. Her malady had not gone into remission. The best way to treat it was not clear, and she appropriately sought the advice of several specialists. I was one of them. All of us echoed the others' words of encouragement; there were still several options to be tried that could make inroads against the disease. Yet our words seemed to fall on deaf ears. Then the primary hematologist noted that the woman's anemia was gradually worsening. He increased her dose of erythropoietin, the genetically engineered growth factor that stimulates red blood cell production in the marrow. The treatment increased her hemoglobin a small amount, but even that boost had a profound impact. "I feel better," she said. "I'm still tired, but not like I was, so fatigued that I didn't want to move out of bed in the morning." With that single change in how she felt, she was able to take in our words of

encouragement. It was as if the relentless signals from her tissues, demanding hemoglobin and its gift of oxygen, had crowded
every corner of her mind and allowed no room for hope.

Kindling and sustaining hope depend not only on images that
may be conjured in the mind but also on how those images are
brought into focus or blurred by the ongoing input of nerves
from organs and tissues to the brain.

BELIEF AND THE AUTONOMIC NERVOUS
SYSTEM: AN EXAMPLE FROM ASTHMA

The autonomic nervous system is the part of the brain and spinal
cord that regulates the essential and automatic physiological
functioning of respiration, circulation, and digestion. We do not
consciously regulate these systems. An elegantly designed network sends signals from the central nervous system to the lungs,
heart, blood vessels, and bowel; the nerves carry signals back to
the spinal cord and brain from the viscera. Experiments with
placebos suggest that belief and expectation can exert potent
effects on the autonomic nervous system. One clinical example
of this involves asthma.

Asthma is a disorder marked by excessive contraction of the
smooth muscles around the bronchi; this contraction narrows
the airways and makes it difficult to breathe deeply and obtain
needed oxygen. The desperate movement of air through the

narrowed bronchi causes the characteristic wheezing sounds of asthma. Any sufferer will testify to the panic caused by struggling to breathe through constricted airways.

The relaxation and contraction of the smooth muscle, and thus the opening and closing of our airways, are controlled by the autonomic nervous system. Its nerves lace the smooth muscle surrounding the bronchi. Under normal conditions, the smooth muscles around the bronchi effortlessly contract or relax. In asthma, the signals from the nerves do not respond effectively. After exposure to allergens like pollen, dust, and mites, or to microbes that cause respiratory infection, the smooth muscles clamp down like a trap, and the airways narrow tightly. Key treatments for people with asthma are drugs called bronchodilators, which open up the airways. They contain chemicals that mimic the dilating effects of the autonomic nerves on the smooth muscle.

In a typical placebo experiment on asthma, the patient, breathing comfortably, has detailed measurements taken of her respiration. She breathes into a spirometer, a bellowslike apparatus, and recordings are made of air volumes inhaled and exhaled, as well as contents of the gases oxygen and carbon dioxide exchanged. The oxygen levels in her blood also are quantitated.

The researcher then says, "Before challenging you with an irritant that will trigger your asthma, I am going to give you a powerful bronchodilator that should keep your airways open." The researcher picks up a cartridge containing albuterol, an adrenaline-like bronchodilator that relaxes the bronchial smooth muscle.

He inserts the albuterol cartridge into a device that looks like a small pipe, with a wide stem and a tight-fitting bowl. The patient knows the routine very well, since this is how she self-administers her bronchodilator medication. The researcher snaps the cartridge into the bowl, and the patient is poised with her lips around the stem. The researcher presses down on the pressurized cartridge, a sharp *whoosh* is heard, and albuterol, under pressure from the cartridge, is expelled through the pipe stem. The volunteer deeply inhales it.

Two minutes elapse, giving time for the albuterol to work. The researcher then tells the patient that he will administer an irritant that normally triggers her wheezing. There is wariness in the patient's eyes, but the researcher reiterates that more than adequate doses of bronchodilator will be given. A second cartridge is inserted into the pipelike device. The albuterol is administered, and another set of measurements of pulmonary function is made. Then the patient inhales the irritant. She appears to be breathing effortlessly and, when asked, states that she seems not to have any wheezing or distress due to the irritant. The bronchodilator has prevented the reaction.

The sequence is repeated several times. Each time there is no sign of wheezing. Then the researcher informs the patient that he will syncopate the sequence with a single administration of inert saline followed by the chemical irritant. The patient knew this was coming when she agreed to the experiment. She sits up in the chair, and her face tightens as she inhales the saline. Within a few seconds after inhaling the chemical irritant, she begins to wheeze. Beads of perspiration glisten over her forehead, and her nostrils flare as she works hard to breathe in and out. Measurements are taken. The volumes of air inhaled and exhaled are significantly lower than those with the bronchodilator treatment.

The researcher quickly inserts an albuterol cartridge into the pipelike device, and the patient gratefully inhales the drug. Her wheezing soon abates.

AGAIN, THE SLEIGHT OF HAND

After reversing the asthmatic attack, the researcher obtains another set of measurements, and the patient's airways are now back to baseline. The researcher picks up a cartridge and inserts it into the pipe. The patient follows this action with her eyes. But, without telling her, the researcher again administers a saline placebo. Then the irritant is given.

Remarkably, the patient's breathing does not collapse into desperate spasms of wheezing. Air flows smoothly in and out of her lungs. She seems entirely comfortable. Measurements are taken, and as she inhales and exhales, the bellows of the spirometer move well. The smooth muscles around her airways did not clamp down.

As with the experiments on pain, belief and expectancy in response to the environmental cues caused changes in brain pathways, here altering the output of the autonomic nervous system. These autonomic nerves released chemicals of the adrenaline family that substituted for albuterol; they relaxed smooth muscle and opened the bronchi.

THE AUTONOMIC NERVOUS SYSTEM
IN THE CLINIC

Heart rate, blood pressure, respiration, and digestion, all autonomic functions, are factors in each patient's illness and response to treatment. It is likely that in certain settings, such as recovery from heart attack, the emotional state of the patient makes a major contribution. There are studies that support this reasonable hypothesis. Patients who are hopeful, largely because of their religious faith and their trust in the physician, have a more rapid return to health and a higher rate of survival.

These data are derived from analyzing large numbers of patients. They differ from the experiments detailed above because the many different variables at play in an uncontrolled clinical setting make it harder to prove that the emotion dictated the outcome. One can say with confidence that there was a correlation between hopefulness and recovery, without saying hope was causative.

The complexity increases in cases like Dan Conrad's. He struggled to breathe because of the rapidly growing mass in his chest. That tumor physically compressed airways, as well as causing inflammation and irritation of the surrounding bronchi. As a scientist, I would not posit that the effects of belief and

expectancy on the autonomic nervous system would be sufficient per se to counteract the constricting effects of his cancer;
the contributions of hope, though, should not be entirely dis-
counted. They are likely to be a smaller fraction than the effects of chemotherapy and radiation on opening Dan's airways.
Nonetheless, we might also reasonably speculate that once the
pressure and inflammation from the lymphoma began to be alle-
viated, hope may have modulated his autonomic system and
added to his recovery.

BELIEF AND THE MOVEMENT
OF VOLUNTARY MUSCLES

While the autonomic nervous system controls involuntary
smooth muscles like those that open and close our airways or
constrict and relax our stomach and bowel, there is another neural network that controls voluntary muscles. Parts of the cortex
of the brain initiate our intention to lift an arm, get up from a
chair, or smile. These nerves course down through the brain,
connecting at multiple points via the spinal cord, and exit to the
specific muscles required for the action.

In August 2001, Dr. A. Jon Stoessl and coworkers at the University of British Columbia in Vancouver reported a study of
placebos and Parkinson's disease in the prestigious journal *Science.* One of the most debilitating aspects of Parkinson's is the

progressive loss of voluntary muscle function. The researchers showed that placebos had an impact on the disorder. Their work indicates that belief and expectation can alter neurochemistry that affects voluntary muscle regulation.

Parkinson's disease comes about because of degeneration of brain nerve cells that produce dopamine. Dopamine is a key chemical controlling motor function, particularly the intentional use of the face, arms, and legs. In the early stages of the disease, there is often a fine tremor of the hands and a subtle reduction in the range of facial expression. As more dopamine is lost, the patient finds it increasingly difficult to walk without shuffling, to write smoothly, to laugh heartily at a joke. The term applied to the change in mien is "masklike face." Late in the disease, the muscles of the arms and legs contract, and the person feels imprisoned by rigid muscles. Boosting the level of dopamine with certain drugs can improve the symptoms by unclamping the rigid muscles and permitting better control of fine movements.

The neurologists in Vancouver used positron emission to-mography (PET), a sophisticated type of brain scan, to assess dopamine activity within the diseased parts of the brain of six Parkinson's patients, both before and after they were injected with either a placebo or a drug that mimics dopamine. PET scans are read on a computer, and the changes in brain activity are reflected as changes in color on the screen. At baseline, before any treatment, the Parkinson's patients' basal ganglia and other dopamine-rich areas of the brain appeared as a pastel blue on the computer screen. The drug that releases dopamine caused these circuits to light up as a glowing orange. As expected, the patients' muscles functioned more fluidly. Next, when the patients who believed and expected that they were receiving the drug were instead given a placebo, the same intensity and type of color change occurred. Placebos caused the brain to release as much dopamine as the active drug. Moreover, the patients had similar objective improvements in movement of their contracted muscles with the placebo.

MAKING CONNECTIONS

Although we have not mapped all the circuits that may account for the remarkable effects of placebos on physiological functions, there are, as Bruce Cohen pointed out, reasonable links that can be drawn. The "reward circuits" that Cohen is studying for their relevance to drug addiction are hypothesized to play a part in the placebo effect. For example, Stoessl in Vancouver speculates that patients with high levels of motivation and goal-seeking behavior process their expectation of benefit as a form of reward. The reward circuits found in the frontal lobes of the brain are rich in dopamine. It is possible but not yet proven that these reward circuits account for the release of the chemical in the frontal lobes and in the deeper parts of the brain, basal ganglia, that are diseased in people with Parkinson's.

Similarly, very recent studies suggest that neural circuits that release dopamine may connect with pathways that provide endorphins and enkephalins to the regions of the central nervous system responsible for relaying pain signals.

THE SPECTRUM OF RESPONSE

It should be emphasized that not every subject in an experiment on placebos or a clinical trial with sham therapy responds equally. Some individuals experience profound effects, others more moderate reactions, and still others minimal effect or none. As with nearly all of human biology, variability reflects the contributions of genetic factors and environmental aspects. Only very recently have studies begun to identify genetic differences that could account for some of the placebo response. For example, the amounts of endorphins and enkephalins released after a measured painful stimulus can vary according to the type of gene that contributes to dopamine metabolism in the brain. Some people have a form of the gene that is correlated with the robust release of endorphins and enkephalins, and thus seem less pained by noxious stimuli. Others with a different form of the gene perceive pain more profoundly. More work needs to be done on this preliminary observation in the context of how we respond to placebos.

Environment also has a major influence on neural circuits. Experiences cause structural changes in the brain, sculpting synapses in profound ways. This "plasticity" of the brain has been demonstrated by neuroscientists over the past decades, most notably by Dr. Eric Kandel of Columbia University, who received a Nobel Prize for his work. Our brains are not set in their structure by the genes we inherit but are continuously molded during the course of our lives. How past and present environments may shape circuits that connect complex emotions (like belief and expectation) with the pathways that release dopamine, endorphins, and enkephalins, or autonomic mediators, is not known. But this dynamic biology emphasizes that we are not prisoners of our DNA. We will likely discover genes that contribute to the very complex feeling we know as hope, but the circuits in the brain that stem from this feeling are not static. Rather, events in our lives modify them, and I would posit that the words spoken and the gestures made by physicians and surgeons and nurses and social workers and psychologists and psychiatrists, and family and friends, influence the synaptic connections. No one should underestimate the complexity of factors that coalesce in this biological process. But I interpret it to mean that no one is beyond the capacity to hope.

As I pondered the complexity of the biology, it seemed time to learn more about positive emotions beyond the placebo effect. I called Dr. Steven Hyman, who was then the director of the National Institute of Mental Health. In that capacity, Steve fostered research on mind and brain across the United States. He directed me to Madison, Wisconsin. I arrived in the midst of a winter snowstorm.

CHAPTER 8

Deconstructing Hope

It was evening when I arrived in Madison to meet Professor Richard Davidson. Sharp crosscurrents of wind off the surrounding lakes cut deeply through the streets as I waited for him. Davidson is an experimental psychologist and one of the world's experts on the biology of positive emotions. We planned to discuss his work over dinner, and the next day I would observe several ongoing experiments.

A cranberry-red vintage Subaru station wagon with the license plate EMOTE pulled up to the curb. A wiry, compact man with a thick head of curly black hair tinged with gray jumped out. "Richie Davidson," he said brightly, "so good to meet you."

As we drove to the restaurant, I asked Davidson why he had devoted his career to studying the biology of joy, resilience, and motivation.

"I was a child of the sixties," he replied, his face widening into a wry smile. "I was deeply affected by Norman O. Brown's *Life Against Death* and *Love's Body*, and by the writings of Herbert Marcuse. As a teenager, I believed we were on the verge of transforming ourselves into new kinds of people building a new kind of society through the power of emotions like love."

While he has relinquished some of his youthful utopianism, Davidson retains the sensibility of an idealist. The aim of his work is to understand how the brain generates enabling emotions and how we tackle life's challenges. This knowledge, he believes, will one day contribute to the development of techniques and treatments that will help the many people in need.

A PSYCHOLOGIST'S WORKING
DEFINITION OF HOPE

I told Davidson that one of the difficulties I experienced in trying to learn about the biology of emotions was the definition of terms. Even "emotion" was not uniformly applied by researchers in the field.

He acknowledged the problem as we discussed it over dinner. Nonetheless, he said, one could effectively communicate data about an emotion by providing a working definition within the context of a specific psychological experiment.

In vain, I had searched the experimental literature for a working definition of "hope." How would he, as an experimental psychologist, deconstruct it?

"I understand hope as an emotion made up of two parts: a cognitive part and an affective part. When we hope for something, we employ, to some degree, our cognition, marshaling information and data relevant to a desired future event. If, like one of your patients, Jerry, you are suffering with a serious illness and you hope for improvement, even for a cure, you have to generate a different vision of your condition in your mind. That picture is painted in part by assimilating information about the disease and its potential treatments.

"But hope also involves what I would call affective forecasting—that is, the comforting, energizing, elevating *feeling* that you experience when you project in your mind a positive future. This requires the brain to generate a different affective, or feeling, state than the one you are currently in."

We often speak in poetic terms, I noted to Davidson; we are "lifted by" hope, hope "has wings." Certainly, this sense of elevation was apparent at the bedside, as I observed patients, and in my own experience as a patient. When Jim Rainville painted a picture of a very different future for me, my feeling of hope encompassed more than information about a differing diagnosis and a new therapeutic approach—it involved a unique feeling state that was intensely visceral, sensed as a sharp upward shift in mood.

The two components, cognition and feeling, are not separate in the brain but interweave and modify each other. "That is what's

so interesting about hope and other emotions," Davidson continued, "how they can powerfully influence cognition and deliberative
thinking. They impact the machinery of perception, the circuits
that are used to take in and process data and make decisions.

"This is one of Antonio Damasio's most important contributions," Davidson said. Dr. Antonio Damasio is a professor of neurology at the University of Iowa. In addition to his scientific
publications, he is the author of several popular books. I had read
each, as well as several of his research articles. "Emotions can
contribute to decision making in a constructive way. The convenient divide between cognition and emotion that psychologists
have used in the past is an artificial, if not misleading, one."

Damasio used a famous case from the annals of neurology to
drive home the point that emotion is an essential companion to
logical decision making. It is the tale of Phineas Gage, a railroad
foreman who suffered a freakish accident in the summer of 1848.
While he was overseeing the laying of track, a metal spike flew
up and penetrated Gage's skull, just behind his forehead.
Remarkably, the spike passed into the frontal lobes of the brain
but did not kill him. Furthermore, when it was removed, Gage
woke up from his coma. Initially, it was difficult to discern the
effects of the damage to his brain, but then they became clear
from his behavior. The injury left him emotionally bereft as well
as devoid of the capacity to reason. In the ensuing years, other
patients who had parts of their frontal lobes damaged or removed because of trauma, tumors, or as a psychiatric treatment
were found to resemble Gage. Their emotions were shallow, and

they suffered deficits in deliberate reasoning. Damasio argues that such clinical cases show that emotions are our "cognitive guides," fostering the process of logic rather than retarding it. This point is even more dramatically shown in rare individuals who appear to lack fear because of damage to their amygdalae.

FEAR AND HOPE

"Amygdala" is Latin for "almond," and this structure sits deep in our brains. It received its name from early anatomists who observed that it was shaped like the nut. There are two such structures, one on the right side and the other on the left side of the brain. The amygdala is a key part of the pathway that mediates fear. Joseph LeDoux, professor of science at New York University, has done seminal research on rodents to define the circuits that enter and exit the amygdala in response to threat and danger. Much of this work is described in his popular books on the nature of emotions.

The amygdala in humans, as in rats, serves as a clearing-house for the integration of stimuli that evoke fear—a mugger drawing out a switchblade knife, the slithering form of an advancing snake in a forest. Davidson emphasized to me how important LeDoux's work was in linking the experience of fear with physiological changes. The amygdala sends signals that cause familiar bodily responses of fear: a pounding heart, tense contraction of muscles, sweat pouring from the skin. In addition to these perceived changes in our physiology, fear causes the release of a cascade of key hormones: CRH, from the lower part of the brain called the "hypothalamus," which acts on the pituitary gland to release another hormone, ACTH, which travels to the adrenal glands to release the stress hormone cortisol. All of this happens within seconds of seeing the flash of the mugger's knife or hearing the hiss of the snake.

It turns out there are rare individuals with damage to brain pathways who appear to be fearless. Some of these patients have been studied by Antonio Damasio, and his results further support the model of emotions as necessary for appropriate decision making.

The case of a young woman, S.M., with the rare, inherited Urbach-Wiethe disease was reported by Damasio's research team in the journal *Nature*. Her malady caused abnormal deposits of calcium to accumulate in her throat, skin, and brain. But the calcium was not randomly speckled; rather, it was deposited selectively in both her left and right amygdalae. As the calcium accumulated in large amounts, it destroyed the cells, so she essentially lived without this part of her brain.

The young woman was extremely affable. Her past history of relationships indicated that she made friends easily and had had several romantic attachments. The drawback was that S.M. had often been taken advantage of by people whom she trusted. Her outgoing personality and readiness to engage others became glaringly apparent in Damasio's research laboratory. Shortly after being introduced to his scientific team, S.M. hugged and touched the researchers. Such intimate physical contact was not socially appropriate within minutes of being introduced to strangers, even though they were vouched for as responsible, professional individuals.

Damasio hypothesized that damage to S.M.'s amygdala had erased her emotion of fear, and that countervailing positive emotions therefore occurred at greater frequency and greater intensity. To experimentally address this hypothesis required more than eliciting information about S.M.'s past attachments or observing her hugging his laboratory researchers.

One way that experimental neurologists and psychologists assess the biology of emotions is by testing people's responses to stereotyped facial expressions. S.M. and several other patients with similar damage to both amygdalae could not recognize the expression of fear in another person's face, though they could easily discern other emotions, like surprise, that can be expressed with similar configurations of facial muscles. Moreover, S.M., who had good drawing skills, could not sketch a face representing fear, although she could draw faces depicting other emotions. When asked to mimic an expression of fear with her own face, she was unable to summon the wide-eyed, tense mien that all of us associate with terror. Finally, when placed in a situation that would normally induce fear—like the threat of a painful electrical shock—S.M. felt no such emotion. On an intellectual level, she knew what fear was supposed to be, what caused it, and what to do in situations that were threatening, yet none of this knowledge appeared to guide her behavior.

Indeed, the lack of fear in people without properly functioning amygdalae is evident in their other social behaviors. For
example, they lack judgment of trustworthiness. This was demonstrated through psychological testing that also used stereotyped facial expressions. When S.M. and similar patients viewed
faces that would be characterized as trustworthy, they correctly
classified them as belonging to people who might be approached
when in need. But they were unable to assess faces that were suspicious, those of people one would ordinarily flee from. In fact,
they judged such threatening faces just as trustworthy as the others. This inability to discern trustworthiness was observed only
in people whose right and left amygdalae were both damaged. In
patients who had one dysfunctional amygdala, the ability to dis-
criminate between trustworthy and suspicious faces was preserved. These experiments reveal how primitive emotions like
fear are transmuted into more sophisticated social behaviors that
require decision making and govern our relationships with others in the workplace and family.

When I read about S.M., I thought about true hope versus
false hope. False hope does not recognize the risks and dangers that true hope does. False hope can lead to intemperate
choices and flawed decision making. True hope takes into account the real threats that exist and seeks to navigate the best
path around them.

Hope, Davidson continued, does not cast a veil over perception and thought. In this way, it is different from blind optimism:
It brings reality into sharp focus. In the setting of illness, hope

helps us weigh highly charged and often frightening information about the malady and its therapies. Hope incorporates fear into the process of rational deliberation and tempers it so we can think and choose without panic. On the other hand, unbridled fear overwhelms initial hope like a tidal wave, blocking the cognitive intake of information and washing away other feelings. Dan Conrad exemplified this attribute.

Davidson's deconstruction of hope and reasoning also seemed applicable to George Griffin. He had as sharp and clear a view of his condition as anyone. So did Barbara Wilson. For both, faith was the wellspring of hope that helped dampen their fear. Each soberly made decisions that were beneficial to body and to spirit.

Hope, then, is the ballast that keeps us steady, that recognizes where along the path are the dangers and pitfalls that can throw us off; hope tempers fear so we can recognize dangers and then bypass or endure them.

CONNECTIONS BETWEEN POSITIVE
AND NEGATIVE EMOTIONS

"One of the most intriguing questions is how we turn off negative emotions," Davidson told me. It was a reasonable hypothesis, based on the data he was gathering, that there are circuits from the prefrontal cortex associated with positive emotions that are inhibitory to regions of the amygdala. "And why some people seem more able to dampen fear by summoning courage, or to overcome despair by holding on to love." Again, this ability is likely to reflect our genes and our life experiences, which continuously sculpt the inherited circuitry of the brain. This is the yin and yang of emotion, the penetration of the positive into the negative, and the negative into the positive.

That interplay, Davidson pointed out, occurs when we set a goal and pursue a path to achieve it. Such behavior seems to involve the reward circuits—rich in dopamine from the prefrontal cortex—that run deep into lower areas that include the amygdala and the hippocampus.

These goal-seeking and reward circuits, Davidson speculated—as had Stoessl with Parkinson's—instantiated a sense of hope in people. Instilling hope in the brain involves setting a firm goal and anticipating the reward of living with the dream fulfilled.

Of course, Davidson emphasized, there is no single "hope center" or "hope neurotransmitter" in the brain, no one region or molecule that itself mediates this emotion. Such a notion amounts to phrenology.* Modern scientists, loath to think like phrenologists, emphasize that there are no sharply circumscribed regions of the brain, each responsible for a single mental process.

---

*Phrenology is a vestige of Victorian science. In the nineteenth century, a popular school of thought held that the bumps of the skull reflected the greater or lesser mass of brain tissue underneath, and that a person's character could be discerned by palpating the bumps. Phrenology was the pop psychology of the era; not only cognitive faculties, like speech and memory, could be correlated with the contour of the skull, but also personality traits, like benevolence and suavity.

While it is more accurate to depict the brain as an integrated woven mantle, with interlacing circuits of neurons that connect different parts (rather than a patchwork quilt of individual functional pieces bordering one another), it is still correct to assign to discrete regions of the brain important functions like speech, visual perception, and memory.

Thus, using sophisticated brain scans to map circuits and the nodes that integrate and process neural signals, Davidson and his colleagues are looking for other key regions. One appears to be in the brain's cortex, near the temples, and is called the "anterior cingulate." The researchers believe that this region participates in the complex process of conflict resolution, the weighing of opposing choices, and the ultimate decision to follow a certain path, a key in how hope may influence cognition.

Davidson pointed out that another element in summoning and sustaining hope is memory. Hope integrates information and feelings derived from present circumstances, and it also draws on experiences of the past, seeking models and directions from other individuals who endured and transcended harrowing situations and overcame seemingly long odds. Again, Dan Conrad came to my mind, his stark memory of Tom Kane in the ICU.

How are memories retrieved? The part of the brain called the "hippocampus" is believed to be integral in this process. This region of the cerebrum has a broad S-shaped sweep; its elegant curvature reminded classical anatomists of a seahorse, so it was given the Greek name for that creature. One type of memory that the hippocampus mediates is "declarative memory," which we experience when we consciously reach back in our minds for previous experiences. The hippocampus also appears to contribute to the linking of objects and events around us with past experiences.

One of the intriguing but still very controversial findings about the role of the hippocampus in memory relates to post-traumatic stress disorder. There is a body of work suggesting that the hippocampus shrinks in people who undergo a life catastrophe. How might this occur? The hippocampus appears particularly sensitive to the stress hormone cortisol. The hypothesis is that people with severe post-traumatic stress disorder have sustained, excessively high levels of cortisol that relentlessly drive the nerve cells of the hippocampus to what might be termed "death from exhaustion." As they lose neurons in the hippocampus, these patients may be impaired in their ability to recall memories that contribute to the generation of positive emotions. An alternative hypothesis, based on studying identical twins, one of whom experienced trauma and the other of whom did not, is that a small hippocampus indicates predisposition to post-traumatic disorder, rather than being the results of the stress. More study is needed to understand the role of the hippocampus in our response to severe trauma.

Patients with post-traumatic stress disorder—like victims of torture, war, rape, and other forms of emotional and physical abuse—are at one end of a spectrum. But there are many others whose lives are deeply changed by a crushing event that is not on the level of Auschwitz or Bosnia. These are people who suffer the loss of a loved one or failure in their career, or debility from an accident or illness. And yet they go on and rebuild their lives. To do so, I said to Davidson, must involve reclaiming hope.

THE LATER-LIFE RESILIENCE STUDY

"Yes. Experimentally, we approach this by studying resilience," he replied. He defines "resilience" as the maintenance of high levels of positive feelings and well-being in the face of significant adversity. It is not that resilient individuals never experience or demonstrate negative feelings; rather, the negative feelings do not persist. He is studying cohorts of elderly Wisconsin residents to identify the roots of resilience.

One study began in 1957 with one third of the high school graduates in the state of Wisconsin. Some ten thousand Wisconsin men and women are participating in this unique medical and psychological study. They are followed over the course of decades, from high school graduation to their advancing years. Remarkably, 97 percent of the cohort has remained in the state. The highly detailed database on these Wisconsin residents is regularly updated with information about their health status and life events. Since the majority of Wisconsin citizens are white and Christian, the group is largely racially and religiously homogeneous, but there is nonetheless diversity in other demographic factors, such as socioeconomic status, rural or urban environment, and education. Out of the group of ten thousand, Davidson has selected a representative cohort sample of five hundred elderly women whose mean age is now seventy-five. He is analyzing their responses to the study's comprehensive psychological questionnaire, and developing a profile of each woman with respect to her affective style. Resilience is a major focus. Blood samples are drawn to provide DNA for future genetic mapping. Each woman then enters Davidson's laboratory and is exposed to specific emotional stimuli. Using EEGs and sophisticated scans, he studies patterns of brain function in response to memories of painful moments in the past. He also presents the women with measured-current stimuli that elicit specific cerebral responses—for example, a sharp, harsh noise that startles them. With changes in the brain activity, measurements are made to correlate activation of particular neural circuits with release of the stress hormone cortisol.

To quantitate the women's handling of stress, Davidson and his coworkers assessed changes in the amount of cortisol, establishing its baseline levels in a home setting, and then in the laboratory during experiments in which the subjects were asked to recall a negative experience. Those with a resilient affect had normal baseline hormone cortisol levels over the course of the day; the hormone was high in the early morning and declined in the late afternoon and evening. When a negative experience was recalled, these "resilient" women did not show a sharp rise in cortisol; this indicated that they would physiologically modulate their response to severe stress.

In contrast to the resilient women were those whom Davidson terms "vulnerable." While recalling traumatic experiences, they had a markedly different cortisol profile: Instead of a sharp downward slope in the hormone, there was a broad, flat plateau. "This difference in regulation of a key stress hormone is significant between the two groups," Davidson said. "And if you look under the curves, meaning sum up the amount of cortisol that was released and worked on both the brain and bodily tissues, it was significantly higher in the nonresilient, vulnerable women, compared to the resilient ones." Furthermore, under laboratory conditions, resilient individuals are able to more quickly extinguish a startle reflex, which prior research has shown to correlate with pathways that connect to the amygdala; this may indicate that women can dampen the fear response.

Davidson is studying the ways in which the bodily health of these women may be modulated by their personalities. He emphasizes that there should be no glib assertions, such as those in the popular media, that the power of the mind is limitless in influencing the clinical outcomes of virtually all diseases.

"From the detailed health histories and current medical status on all of these Wisconsin women, we hope to be able to address how different affective styles, like resilience, may play out over life in terms of their health." Davidson expects that the emotional differences among individuals will not be a sole or a primary determinant but one of several factors that contribute to well-being or illness. For example, since it was winter, the women in his study had recently received the flu vaccine. Those who responded best to the vaccine by producing the largest amounts of antibody against influenza were the ones who had a resilient affect, while the women who would be characterized as vulnerable mustered markedly lower amounts of antibody against the same vaccine. Cortisol can also modulate the immune system, particularly the interplay of different lymphocytes that produce antibodies.

"It's an intriguing observation," Davidson said, "and would prompt me to believe that the state of mind might be important in contributing to the immune response—*at this level*. But I would be very wary of jumping to global conclusions from such data about serious diseases and the immune system."

Some of the papers I had reviewed in my own area tried to draw correlations between affect and the outcome of HIV infection. They were unconvincing. A trait like resilience can be vital to enduring the vicissitudes of AIDS. But in the face of such a destructive virus, which has the force of a megaton bomb on the lymphocytes of the immune system, the fluxes in cortisol level that might occur in individuals who coped well, or did not cope well, with their condition were unlikely to be of a magnitude to influence the clinical outcome in a significant way.

Davidson and other serious researchers chipping away at the large, complex, multifaceted processes that we know as hope readily admit that we are at the cusp of understanding, and that we will need decades of deeper exploration to unravel the biology. But the tools and technology to explore are gradually coming to hand. Circuits involved in positive and negative emotions and their links with stress hormones like cortisol are being mapped. As the genes that code for the components of these neural circuits and their chemical transmitters are identified, scientists can reach into the comprehensive database being built from cohorts like the Wisconsin women, assess the contributions of heredity and life experience to the biology of hope, and better understand why it is so essential to living.

The wisdom of modern science is working to catch up with the wisdom of the ancient Greeks.

Lessons Learned

It has been nearly thirty years since I walked into Esther Weinberg's room. Now, when I meet a new patient, listen to his history, perform a physical examination, review his laboratory tests, and study his X rays, I am doing more than gathering and analyzing clinical data. I am searching for hope. Hope, I have come to believe, is as vital to our lives as the very oxygen that we breathe.

My focus beyond the strictly clinical science was widened by the many patients whom I cared for during those three decades—patients like Frances Walker and Dan Conrad and Barbara Wilson. I am sure there will be others who will teach me more about hope. While my own research continues to concentrate on understanding why cancer cells grow and spread in an unchecked way, how viruses like HIV and hepatitis C invade and destroy healthy cells, and what guides the genesis of blood in the bone marrow, today I follow with intense interest experiments on emotions and how the brain and the body biologically talk to each other.

The question—why some people find and hold on to hope while others do not—was what moved me to write this book. There is no one simple answer. But insights can be drawn from the experiences recounted here. For many who cannot see hope, their vision is blurred because they believe they are unable to exert any level of control over their circumstances. When I encounter such patients, like Esther Weinberg, I try to discover why they feel so completely at the mercy of the forces around them. Helping them find hope harks back to the two elements of the emotion that Richard Davidson discussed: a cognitive, or "informational" one, and an affective, or "feeling" one.

I learned that it takes much more than mere words to communicate information and to alter affect. Dan Conrad forcefully showed me how often I fall short of the mark on both counts. I try hard to let patients read in my eyes that there is true hope for them. The distillation of Sharon Walker's words comes to mind: that for a physician to effectively impart real hope, he has to believe in it himself. Even then, the stated facts and the chance of prevailing are often best articulated in more than one voice. Doctors are fallible, not only in how they wield a scalpel or prescribe a drug but in the language they use. Family and friends, clergy and social workers, psychologists and counselors, and, often most compellingly, other patients like Dotty Hirschberg and George Griffin can better speak from personal experience and reach the roots of despair and distrust.

While it is a convenient construct to divide hope into a cognitive and an affective component, the two are tightly coupled. Feelings and emotions mold logical thinking and deliberate decision making, as Damasio's studies of patients like S.M. show. True hope, then, is not initiated and sustained by completely erasing the emotions, like fear and anxiety, that are often its greatest obstacles. An equilibrium needs to be established, integrating the genuine threats and dangers that exist into the proposed strategies to subsume them. So when a person tells me that he doesn't want to know about the problems and risks, that he believes ignorance is necessary for bliss, I acknowledge that yes, unbridled fear can shatter a fragile sense of hope. But I assert that he still needs to know a minimum amount of information about his diagnosis and the course of his problem; otherwise, his hope is false, and false hope is an insubstantial foundation upon which to stand and weather the vicissitudes of difficult circumstances. It is only true hope that carries its companions, courage and resilience, through. False hope causes them to ultimately fall by the wayside as reality intervenes and overpowers illusion.

Each disease is uncertain in its outcome, and within that uncertainty, we find real hope, because a tumor has not always read the textbook, and a treatment can have an unexpectedly dramatic impact. This is the great paradox of true hope: Because nothing is absolutely determined, there is not only reason to fear but also reason to hope. And so we must find ways to bridle fear and give greater rein to hope. George Griffin knew this long before I did. And, as Dan Conrad and many others since have seen, the uncertainty of science also brings hope. Science may progress in time to produce new therapies that can make the incurable curable.

Many people searching for hope look to their faith. When confronting suffering and loss in my own life, I have found strength and solace in both the insights of tradition and the structure of ritual. Still, as the great Protestant theologian Paul Tillich noted, true faith does not discount doubt. There are times when I doubt and do not see how faith can bolster hope. At such moments I think back to George Griffin, and especially the psalm that he recited during his struggle. There is deep comfort in the sense that we are not alone when we try to move out of the shadow of death. I also think of Barbara Wilson, whose faith enabled her to sustain the calm and balance to address the yearnings of her soul. There are also instances when patients and their families look directly to God for assistance in a cure. I once asked one of these patients, a middle-aged Italian-American woman with breast cancer who attended mass regularly, what words she addressed to God: "I pray that he helps my doctors, that he gives them wisdom." That has become my prayer.

Patients are awash in a sea of statements about the link between their emotions and their maladies. For years I diverted or dismissed their inquiries because I did not know how to answer. Now my response is formed by the lessons taught to me

Acknowledgments

The idea to write about hope came to me at the end of a long and trying week seeing people in the hospital with serious illness. Walking back to my lab from the wards, I asked myself what, beyond any treatment, I could offer my patients. The answer was hope.

For several days the idea of trying to capture this sentiment in prose rattled around in my head. It was both exhilarating and terrifying. Maurice Lamm, who wrote a book about his daughter's illness, *The Power of Hope,* opens with these lines: "We know in our bones that hope is everything. In the back of our minds, we suspect that it is nothing at all." I feared that I would fail to authentically represent how vital and nourishing hope could be, and that failure would reinforce the perception of hope as insubstantial.

I realized I had much to learn about hope. More than two millennia ago, Rabbi Yehoshua ben Prachia, pondering how best to gain new knowledge, said, "Make for yourself a teacher and acquire for yourself a friend." In the writing of this book, there has been little distinction between teacher and friend. I am fortunate to have many friends who taught me a great deal. Their insight, guidance, and criticism were key when I so frequently floundered. They also tolerated my exasperating obsessive rewriting, epitomized by my wife, Pam, asking, "Have you spent three hours changing one comma?"

As with all uncertainties, I sought regular reassurance. Much of it was provided by Pam. In addition, she assisted me in examining the lives of my patients and read draft after draft with her merciless eye.

Above all, she pressed me to be brutally honest about my ignorance and limitations. Pam was integral to the creation of this book.

Youngsun Jung, my assistant of some eighteen years, lent not only her extraordinary word processing skills but also her perspective as a woman of deep Christian faith to the crafting of the manuscript. Her contribution cannot be overstated.

Suzanne Gluck, my agent at William Morris, greatly refined the concept of the book and worked with her characteristic intensity and bluntness in helping me take an amorphous idea and sharpen it into a coherent pursuit. No author could be more fortunate than to have Suzanne. Her assistant, Emily Nurkin, was always available, acting in her typically effective way.

Ann Godoff and Meredith Blum, then at Random House, were extraordinary editors. Early on, before taking pen to paper, I aired my misgivings with Ann about how hard it seemed to represent hope in a narrative form. Ann said that if the book were to succeed, everything that was written had to resonate as authentic. One potential pitfall was to veer into banality or cliché, and Ann promised me that she would be on "schmaltz patrol," meaning that she would prevent me from lapsing into the maudlin. Meredith brought deep insight into how to structure the book and into the meanings that could be drawn from the stories of my patients and myself. Her wisdom is all the more startling given her young age. When Ann and Meredith left Random House, I was lucky to have Jon Karp as my editor to carry the book forward with his signature energy and enthusiasm, and to have the continued commitment of Gina Centrello and her team.

As for old and new friends who proved to be teachers, some live literary lives as writers: Sarah Elizabeth Button White, Marjorie Williams, Tim Noah, Doreen Carvajal, Keith Johnson, Stuart Schoffman, Jane Praeger. Others, who have a keen literary sensibility but are not professionals, served as critical readers: Liz Young, Gerald Menke, Arthur Cohen, Francine and Harry Hartzband, Muriel Pollet, Eric Baker, Everett Fahy, Rick Rawlins. And there were friends who offered much needed encouragement: Frank Rich, Alex Witchel, Margo Howard, Abe and Cindy Steinberger, Michael Share, Anthony Rao. Both Professor Richard Davidson and Dr. Ted Kaptchuk kindly reviewed the scientific accuracy of the chapters. Dr. Steven Hyman was, as always, terribly generous in educating me about current concepts of the mind and in steering me to the best sources to read. Any errors are mine alone.

Although they were not directly involved in the editing of this book, I owe a deep debt of gratitude to my mentors at *The New Yorker*. Over the past seven years, they have taught me so much about the craft of writing: David Remnick, Dorothy Wickenden, Henry Finder, Tina Brown, Alice Truax, Meghan O'Rourke, Daniel Zalewski. I still have much to learn from all of them. I also have benefited greatly over the years from my interactions with *The New Republic,* specifically from the talent of Marty Peretz, Leon Wieseltier, and Peter Beinart. Michal Ronnen Safdie lent her skilled eye and photographic art.

My greatest teachers, some of whom I count among my friends, have been my patients. Any shortcomings in substance or style reflect my own deficiencies. By opening up their lives to me, they have given me gifts that cannot be adequately described in words.

Notes

8 In the event of a radical mastectomy: Radical mastectomy was standard treatment for breast cancer, based on an incorrect understanding of the biology of the tumor. There was an entrenched belief that removing all surrounding tissue was necessary and would completely eradicate the cancer. Any questioning of radical mastectomy was met by stiff resistance, even as research data were generated that challenged its necessity. An excellent book about the medical aspects of breast cancer as well as the social and political forces around its therapy is Dr. Barron H. Lerner's *The Breast Cancer Wars: Hope, Fear, and the Pursuit of a Cure in Twentieth-Century America* (New York: Oxford University Press, 2001). Currently, women choose from among a variety of options, including simple mastectomy, which preserves the muscles and soft tissue below the breast, and lumpectomy, removal of the tumor and sampling of lymph nodes followed by radiation.

16 *"Ayshes chayil me y'imtza"*: Eishet Hayil, *Siddur Sim Shalom,* The Rabbinical Assembly (New York: The United Synagogue of America, 1985), 725.

20 When Miriam, the sister of Moses, slanders his wife, God makes her leprous: This is recounted in Num. 12.

The following commentary is from the *Etz Hayim Torah,* a recent translation of the Bible (Jewish Publication Society,

2001), 652–653: "Today we recognize that it is medically inaccurate and psychologically cruel to tell someone that he or she is afflicted with illness as punishment for behavior not organically related to the illness, and that failure to heal is to be blamed on a lack of will. It should be noted that the Torah itself presents *tzara·at* (leprosy) as an affliction to be cured, not as a punishment to be explained. We might ask: What actions or conditions cause an individual to be isolated from the community of today? And what can religious institutions do to restore that person to the community?"

26 Hope can arrive only when you recognize that there are real options and that you have genuine choices: This has become apparent to me in the many circumstances where we search for hope, not only during illness. Part of that belief comes from reading the journals and studies of people who have experienced the most extreme of circumstances. Notable among these are Primo Levi, Elie Wiesel, and Natan Sharansky, whose hope was focused not only on survival but on sustaining their dignity in the Nazi death camps and Soviet gulag. This core facet of hope, in my experience, applies as well under less extreme conditions, when we are seeking a better future in our relationships with loved ones, in our economic circumstances, even in our political choices. Hope is linked to a sense of control, both over oneself and over external forces; this was the key lesson taught to me by Esther's tragic case.

CHAPTER 2: FALSE HOPE, TRUE HOPE

29 Frances had Stage D colon cancer, the most advanced stage: Cancers are classified in various ways. The stage refers to the extent of the tumor. For a discussion of commonly used classification systems for colon cancer, see V. T. DeVita, Jr., S. Hellman, and S. A. Rosenberg, eds., *Cancer: Principles and Practice of Oncology,* 6th ed. (Philadelphia: Lippincott Williams & Wilkins, 2001).

Tumors are also classified with regard to the aggressiveness of the cells. The vast majority of cancers of the colon are adenocarcinomas. Their behavior is generally aggressive once they have spread beyond the bowel, although not always. In some cases, if there is a single, well-circumscribed metastasis in one segment of the liver, like the left lobe, it may be surgically removed and patients can have a very prolonged remission. When there are multiple metastases, as in Frances's case, with many lesions within the lobe, it is not possible to effectively remove them by surgery.

For a more detailed discussion of the biology and clinical treatment of colon cancer, see DeVita et al., eds., *Cancer*. In addition, information can be obtained from the National Cancer Institute or the American Cancer Society.

Because Stage D colon cancer is so rarely cured, prevention is key. This is done by testing stool for blood during a regular physical examination, as well as, after the age of fifty, undergoing a colonoscopy. For those individuals who have a family history of colon cancer, or of polyps, which are fleshy growths in the colon that can transform into cancer, colonoscopy should be performed at an earlier age and on a more frequent basis.

55 the tide had turned for previously devastating diseases, such as testicular cancer: The story of the discovery of cisplatin is a fascinating one. Scientists were studying the behavior of bacteria in electric fields by introducing a current through a solution containing the microbes. It turned out that the bacteria were killed during the experiment. Many researchers would just ignore this finding, but not in this case. Extensive study went into trying to learn why the bacteria had died, and it turned out that platinum, a heavy metal, had leached out of the electrodes into the solution. The toxic platinum compound quickly killed the rapidly dividing bacteria. Since cancer cells are also rapidly dividing, it seemed that platinum might be useful as some form of antitumor therapy.

Dr. Lawrence Einhorn of Indiana University is credited with performing the key clinical studies that demonstrated the extraordinary effects of cisplatin in testicular cancer. One of the most famous survivors of testicular cancer, whose story exemplifies the melding of medical treatment and personal will, is Lance Armstrong; see his book *It's Not About the Bike* (New York: Putnum Publishing Group, 2000).

57 Richard had come closer to the middle ground where both truth and hope could reside: This is certainly one of the most complex and important aspects of the art of medicine.

There is no perfect or single way to provide clinical information to all patients under all circumstances—there is no "one size fits all" approach. I still struggle with this after many years of medical practice. In general, I follow the patient's lead; some patients want highly detailed data, while others prefer only general knowledge about the usual course of their malady. Still, I believe there needs to be a minimum of information provided so that patients can plan for all possible outcomes, both bad and good.

Recently, more attention has been paid to teaching medical students and young physicians how to "break bad news." For example, see "Dying Words: How Should Doctors Deliver Bad News?," by Jerome Groopman, in *The New Yorker*, October 28, 2002. This can be accessed on my website: www.jeromegroopman.com.

CHAPTER 3: THE RIGHT TO HOPE

65 Stomach cancer is the number two cause of cancer deaths worldwide: For an excellent discussion of the epidemiology, causative factors, staging classification, and treatment approaches to stomach cancer, see V. T. DeVita, Jr., S. Hellman, and S. A. Rosenberg, eds., *Cancer: Principles and Practice of Oncology*, 6th ed. (Philadelphia: Lippincott Williams & Wilkins, 2001).

H. pylori has also recently become a focus of considerable research. Eradication of the bacterium with antibiotics has been shown to markedly reduce the risk for certain stomach cancers in high-incidence areas of the world like Japan. See N. Uemura, S. Okamoto, S. Yamamoto, N. Matsumura, S. Yamaguchi, M. Yamakido, K. Taniyama, N. Sasaki, and R. J. Schlemper, *"Helicobacter pylori* Infection and the Development of Gastric Cancer," *The New England Journal of Medicine* 345 (2001): 784–789; J. G. Fox, and T. C. Wang, *"Helicobacter pylori*—Not a Good Bug After All" (editorial), *The New England Journal of Medicine* 345 (2001): 829–832; L. T. Chen, J. T. Lin, R. Y. Shyu, C. M. Jan, C. L. Chen, I. P. Chiang, S. M. Liu, I. J. Su, and A. L. Cheng, "Prospective Study of *Helicobacter pylori* Eradication Therapy in Stage I_E High-Grade Mucosa-Associated Lymphoid Tissue Lymphoma of the Stomach." *Journal of Clinical Oncology* 19 (2001): 4245–4251.

76 The Lord is my shepherd: The translation is from the King James Version of the Bible, which is the one George Griffin read.

78 This liturgical insight came from the rabbi of my synagogue, William Hamilton: William Hamilton is the rabbi of Kehillath Israel synagogue in Brookline, Massachusetts.

79 Oliver Wendell Holmes, a nineteenth-century Boston physician: There are many sources about Holmes, a remarkable humanistic physician who was one of the great teachers at Harvard Medical School. For more of his insights, there are excellent compilations on the Web.

CHAPTER 4: STEP BY STEP

82 There was little we could do to inhibit HIV, or to stave off fatal complications for long: There are many sources that provide a sense of the clinical, cultural, and emotional impact of the early days of AIDS. A classic charting of the epidemic is *And the Band Played On: Politics, People, and the AIDS Epidemic,* by Randy Shilts (New York: St. Martin's Press, 1987). Recently, an oral history

has been compiled by Ronald Bayer and Gerald M. Oppen-heimer, *AIDS Doctors: Voices from the Epidemic: An Oral History* (New York: Oxford University Press, 2002).

83 Then, in the early 1990s, my clinical practice began to signifi-cantly change: In addition to retinoids, arsenic-based compounds were found to be particularly effective in the treatment of acute promyelocytic leukemia. See "Chairman Mao's Cure for Cancer," by Elisabeth Rosenthal, in *The New York Times Magazine,* May 6, 2001.

83 The outcome of AIDS also dramatically turned around: This, of course, refers to treatment in countries with the resources to pay for such expensive therapies. The United Nations is spearheading an effort to more effectively intervene in Africa and Asia to pre-vent transmission of HIV through education, condom distribu-tion, and treatment of infected pregnant women; there are different points of view with regard to the feasibility and ethics of drug development in these impoverished regions. For example, see Michael Specter's articles "India's Plague," *The New Yorker,* De-cember 17, 2001, and "The Vaccine: Has the Race to Save Africa from AIDS Put Western Science at Odds with Western Ethics?," *The New Yorker,* February 3, 2003; and Larry Kramer's "The Plague We Can't Escape" (op-ed), *The New York Times,* March 15, 2003.

85 It showed an aggressive type of non-Hodgkin's lymphoma: There are different classifications of lymphoma, but the two major groups are Hodgkin's disease and non-Hodgkin's lymphoma. The classification is based on the appearance of the cells, Hodgkin's disease having the signature Reed-Sternberg cells and non-Hodgkin's lymphoma lacking such signature cells and having its own characteristic morphology. More recently, lymphomas have been classified on a molecular basis with regard to which genes are abnormal. This is the first step toward reclassifying these

diseases based on the alterations in their molecular machinery, particularly which oncogenes and tumor-suppressive genes are aberrant. Initial data indicate that prognosis may be powerfully dictated based on such molecular findings. Ultimately, treatment may be rationally guided to target the underlying genetic changes rather than applying a blanket approach of chemotherapy and radiation.

96 Strictly speaking, he's not depressed: Psychiatrists like Jon Levin practice what is termed liaison psychiatry, intervening with patients in the midst of clinical illness. Depression is a discrete diagnosis based on criteria set forth in the *Diagnostic and Statistical Manual (DSM)*. There is, of course, a vast literature on depression. One good source, from the patient's point of view, is Andrew Solomon's *The Noonday Demon: An Atlas of Depression,* (New York: Scribner, 2001). There are profound physical manifestations of depression that importantly factor in formulating a treatment plan for patients. Integration of the care offered by psychiatrists, psychologists, and psychiatric social workers into "standard" physician treatment cannot be overemphasized with regard to its emotional and physical value.

109 We had given him G-CSF, the protein that boosts white cells: G-CSF, or Neupogen, is one of several naturally occurring proteins that regulate white cell production from stem cells in the bone marrow. Another clinically used growth factor is GM-CSF, granulocyte-macrophage colony-stimulating factor, or Leukine. These white-cell growth factors have proven to be extremely effective in shortening the time period and severity of neutropenia (low white count) following chemotherapy, thereby reducing the risk of life-threatening infections.

115 The nurse attending to Dan that day was Deirdre Dolan: The unsung heroes of clinical care are nurses. During the three decades of my training and practice, I've been repeatedly struck by the

fact that it is the nurse who spends the most time with patients, and often has an extraordinary clinical acumen. During my training as an intern and resident, I was regularly guided by the nursing staff in my care of patients, and that has not changed over the years.

118 The experimental antibody that Dan received was further modified by genetic engineering: Monoclonal antibodies have become key components of the therapeutic armamentarium against lymphoma. Currently, Rituxan (Rituximab) is approved for relapsed or refractory, low-grade or follicular, CD20-positive, B-cell non-Hodgkin's lymphoma, and is being tested widely in a variety of different types of B-cell malignancies. A radio-conjugate, which is a radioactive isotope linked to an antibody, is the essence of Zevalin (ibritumomab tiuxetan). The rate of remission has markedly increased with the use of these monoclonal antibodies, and many experts in the field predict that the long-term cure rate will be similarly increased.

It is of note that monoclonal antibodies were initially ushered in as "magic bullets," with claims that they would cure virtually every type of cancer. This did not occur. They then went through a long period languishing as therapies; some were even dismissed as never having been of value. Now, they have reemerged through meticulous clinical testing as key treatments for certain cancers. This "roller coaster" of hype, despair, and then appropriate hope for an experimental treatment is typical for many of the cutting-edge products coming from research laboratories and pharmaceutical companies around the world.

119 Dan's sense that he would die was articulated in vague and visceral terms: More extensive discussion of what may be termed "body-mind" connection whereby input from tissues and organs potently influences our emotional state is in chapter 7, "The Biology of Hope."

120 She purposely positioned Dan next to Dotty Hirschberg: Deirdre
Dolan's intervention is one that I now adopt in my clinical practice. Respecting the boundaries of confidentiality, I will seek out
a patient and ask that patient's permission to put him or her in
contact with a person with a similar medical problem. I also
strongly encourage my patients to attend professionally run support groups, where they learn more about their malady and share
the emotional experience of their illness in a format that fosters
hope.

CHAPTER 5: UNDYING HOPE

122 The dogma that a radical mastectomy: See note to chapter 1,
page 8.

123 She lingered on a pastel drawing of a rainbow: The rainbow
appears in Gen. 9:12–17. I often look at the drawing with the
verses from the Bible and wonder why the ancient Hebrews took
this natural phenomenon as a sign of life's continuity. Light, of
course, is the first act of creation in the Genesis narrative. The
rainbow emanating from sunlight must have inspired great awe in
ancient peoples who contemplated the natural world. One modern reading of the text emphasizes that the appearance of the
rainbow might spark not only awe but also questioning—that is,
how can white light be transformed into its colored component
parts? Questioning is the basis of science, and science, to my
mind, is one of our greatest sources of hope.

124 "The tests done on the biopsy tissue show that the breast-cancer
cells have what is called HER-2 protein on their surface": HER-2,
also known as ERB-2/Neu, is an oncogene that is normally
present in healthy tissues and, in about a third of breast cancer
cases, is overtly expressed. There are large quantities of the
HER-2 protein on these breast cancer cells. Recent studies have
shown that treatment with Herceptin plus chemotherapy not

only increases the chance of remission but also can extend life in those who respond. The original monoclonal antibody directed against HER-2 protein was derived from a mouse, and then, through recombinant DNA technology, humanized antibodies were produced. The risk of damage to heart muscle is believed to be due to the presence of HER-2 protein normally found in cardiac tissue.

131 I want to start with Xeloda: Xeloda (capecitabine) is a fluoropyrimidine carbamate with antineoplastic activity. It is an orally administered systemic prodrug of 5'-fluorouridine (5'-DFUR) which is converted to 5-fluorouracil. The chemical name for capecitabine is 5'-deoxy-5-fluoro-N-[(pentyloxy) carbonyl]-cytidine; it has a molecular weight of 359.35.

132 *Doctoring,* by Dr. Eric Cassell: Cassell refers to the very roots of medicine as being intertwined with the concept of love: "On the other hand, in the Hippocratic Corpus it says 'Love of the art starts with love of humankind.'" Eric J. Cassell, *Doctoring: The Nature of Primary Care Medicine* (New York, Oxford University Press, 1997), 183.

141 About three decades ago, Dr. Elisabeth Kübler-Ross: The work of Elisabeth Kübler-Ross was indeed groundbreaking, particularly her first book, *On Death and Dying: What the Dying Have to Teach Doctors, Nurses, Clergy, and Their Own Families* (New York: Macmillan, 1969).

143 But many whose work is primarily with the dying have taken issue with her paradigm: See especially Sherwin Nuland's *How We Die: Reflections on Life's Final Chapter* (New York: Knopf, 1994).

CHAPTER 6: EXITING A LABYRINTH OF PAIN

147 When the pain did not immediately subside, I underwent a discectomy: For a discussion of decision making around low back pain, see "A Knife in the Back," by Jerome Groopman, *The New Yorker,* April 8, 2002; this can be accessed on my website, www.

jeromegroopman.com. One view on the contribution of psychological distress to somatization expressed as low back pain is John Sarno's *Healing Back Pain: The Mind-Body Connection* (New York: Warner Books, 1991). A recent clinical review on the subject, by Richard A. Deyo and James N. Weinstein, is "Low Back Pain," *The New England Journal of Medicine* 344 (2001): 363–370.

146 X rays showed no clear cause for the pain: A significant fraction of patients have no clear anatomical explanation for their low back pain. Nearly all patients have so-called degenerative changes that include narrowing of the space between vertebrae or some bulging of the discs, but whether such changes contribute to back pain in any particular case is difficult to discern. Similarly, a third or more of healthy middle-aged individuals with no back pain symptoms will have evidence for a torn annulus or ruptured disc on a CAT scan or an MRI, so that even the presence of such findings does not necessarily account for the pain.

CHAPTER 7: THE BIOLOGY OF HOPE

163 The mind is a product of the brain: For a detailed discussion of connections between brain and body, both neural and hormonal, see Esther M. Sternberg and Philip W. Gold, "The Mind-Body Interaction in Disease," *Scientific American,* Special Edition: The Hidden Mind, 12 (2002); Esther Sternberg, *The Balance Within: The Science Connecting Health and Emotions* (New York: W. H. Freeman & Co., 2000).

163 Stress was agreed to be a bad thing: For a more in-depth discussion of stress, stress alleviation, and stress's impact on cardiovascular function, see Herbert Benson's *The Relaxation Response* (New York: Wholecare, 2000), as well as Bruce S. McEwen's article "Protective and Damaging Effects of Stress Mediators," in *The New England Journal of Medicine* 338 (1998): 171–179.

164 McLean Hospital, one of the preeminent institutions for psychiatric care: The colorful history of the hospital and its place in

American society is recounted by Alex Beam in *Gracefully Insane: The Rise and Fall of America's Premier Mental Hospital* (New York: PublicAffairs, 2001).

167 "'Mind' is a manifestation of brain": An accessible recent popular book describing neurochemical substrates and connections that contribute to perception and thinking is Steven Pinker's *How the Mind Works* (New York: W. W. Norton, 1997).

168 The term "placebo": There are several excellent sources for further reading on the history of placebos and the use of placebos in clinical medicine: Anne Harrington, *The Placebo Effect: An Interdisciplinary Exploration* (Cambridge, Mass.: Harvard University Press, 1997); Walter Brown, "The Placebo Effect," *Scientific American*, January 1998; Howard Spiro, *The Power of Hope: A Doctor's Perspective* (New Haven, Conn.: Yale University Press, 1998); Ted J. Kaptchuk, "Powerful Placebo: The Dark Side of the Randomized Controlled Trial," *Lancet* 351 (1998): 1722–1725.

A controversy over placebo effects was sparked by an article by two Danish researchers, Asbjorn Hrobjartsson and Peter Gotzsche, "Is the Placebo Powerless? An Analysis of Clinical Trials Comparing Placebo with No Treatment," *The New England Journal of Medicine* 344 (2001): 1594–1602. This was a retrospective analysis of clinical trials and emphasized that it would be useful to have a "no treatment" cohort in any clinical study that compared active treatment with placebo. As Walter Brown writes in *Nature* (422 [2003]: 118–119), there were flaws in the meta-analysis by Hrobjartsson and Gotzsche. But the controversy is a healthy one since current efforts are better focused on defining the depth and extent of the placebo effect in various medical conditions using more rigorous experimental design.

One topic outside the scope of this book that is important in the placebo effect relates to psychological maladies, particularly depression. Some studies indicate that the placebo effect may account for some 30 percent or more of the clinical benefit of antidepressive medicines. See Walter Brown's discussion in *Scientific American* (January 1998). For a "cinematic" depiction of this phenomenon, see the PBS special with Alan Alda, featuring a recent study at the University of California, Los Angeles, on patients with depression: PBS's *Scientific American Frontiers*, "The Wonder Pill," www.pbs.org/saf/1307.

The flip side of the positive effects of placebo is the so-called nocebo effect. This term is used to describe adverse effects or apparent toxicities from what are chemically inert treatments. This is a fascinating area of study and one that shows how negative beliefs and negative expectations can affect perception and certain aspects of the body's physiology. This is reviewed by Drs. Arthur J. Barsky, Ralph Saintfort, Malcolm P. Rogers, and Jonathan F. Borus in "Nonspecific Medication Side Effects and the Nocebo Phenomenon," *The Journal of the American Medical Association* 287 (2002): 622–627. There are a number of fascinating experiments described in the article. For example, the power of suggestion with regard to development of symptoms was dramatically demonstrated in experiments in which normal volunteers had an electrode placed on each side of the head, after which they were told that a low-voltage electric current would be passed across the electrodes. The researcher turned a dial that supposedly produced the electric current, but in fact, it was a "dummy" dial and no current whatsoever was passed. A significant proportion, between a third and a half, of normal healthy subjects reported symptoms such as headache coincident with the nonapplication of the "electric current" across their heads. Similarly, patients participating in placebo-controlled trials who received placebo often report symptoms such as nausea, headache, muscle pain, feverish feelings, etc. In the PBS documentary cited above, normal young healthy subjects participating in a study of herbal medication for the common cold versus placebo were interviewed and asked whether they believed they were receiving the active treatment or the placebo. A significant proportion of individuals on placebo believed they were receiving active therapy because of "side effects" that they felt upon initiating the therapy, although the therapy was entirely inert.

From a clinical point of view, the nocebo effect can be a most vexing problem. One patient, for example, said to me, "I don't want to know about possible side effects of the drugs you are prescribing, since I always get the side effects whenever they are told to me." The issue is that some of the side effects are very important to know about, since they may indicate an early allergy to the medication that, if continued, could cause damage to skin, eyes, etc.; other side effects of the treatment, such as nausea, could be the earliest signs of toxicity to vital organs like the liver. I explain this to patients, and state that I prefer that they "cry wolf," meaning report any and all symptoms even if these prove not to be significant, or have been "suggested," in lieu of not reporting symptoms since they are not aware of them. Ignorance puts them at risk to suffer potentially preventable clinical problems.

169 To understand the experiments on placebos and pain: The National Institutes of Health in Bethesda, Maryland, convened an interdisciplinary conference on "The Science of the Placebo," November 19–21, 2000. This was an extraordinary gathering of researchers and clinicians from across the spectrum, ranging from neurophysiologists to anthropologists to statisticians to complementary medical practitioners. Dr. Donald D. Price at the University of Florida, Gainesville, and Dr. Lene Vase Soerensen from the University of Aarhus, Denmark, contributed a paper, "Biological and Psychological Mechanisms of Placebo Analgesia," to this gathering. A lucid and detailed description of pain pathways and neurotransmitters can be found in that paper.

A review of the placebo effect in antidepressant therapy can be found in a meta-analysis by Drs. Irving Kirsch and Guy

Sapirstein, "Listening to Prozac but Hearing Placebo: A Meta-analysis of Antidepressant Medication," *Prevention & Treatment* 1 (1998), article 2a.

170 Belief and expectation, cardinal components of hope: Researchers have focused on other aspects of placebo effect beyond belief and expectation, specifically classical conditioning à la Pavlov's experiments with dogs. For a discussion of how we may be conditioned by prior experiences and how this is a component of the placebo effect, see G. H. Montgomery, and I. Kirsch, "Classical Conditioning and the Placebo Effect," *Pain* 72 (1997): 107–113.

Recently, brain imaging technology, including positron emission tomography, has been used to trace neural networks in the brain that mediate opioid analgesia and placebo analgesia. A recent study from the Karolinska Institute in Stockholm and the Helsinki University Hospital in Finland showed that a related neuromechanism appears to mediate both placebo and opioid analgesia: see P. Petrovic, E. Kalso, K. M. Petersson, and M. Ingvar, "Placebo and Opioid Analgesia—Imaging a Shared Neural Network," *Sciencexpress,* February 7, 2002.

With regard to how cerebral circuits in depressed subjects may be altered with placebo treatment, see recent work from A. F. Leuchter, I. A. Cook, E. A. Witte, M. Morgan, and M. Abrams, "Changes in Brain Function of Depressed Subjects During Treatment with Placebo," *American Journal of Psychiatry* 159 (2002): 122–129. Their study indicates that "effective" placebo treatment induces changes in brain function that are distinct from those associated with antidepressant medication.

An in-depth discussion of stress is to be found in Bruce S. McEwen's article "Protective and Damaging Effects of Stress Mediators," in *The New England Journal of Medicine* 338 (1998): 171–179. Also see Thomas Elbert and Brigitte Rockstroh, "Stress Factors: The Science of Our Flexible Responses to an Unpredictable World," *Nature* 421 (2003): 477–478.

170 A classical Benedetti experiment: M. Amanzio, A. Pollo, G. Maggi, and F. Benedetti, "Response Variability to Analgesics: A Role for Non-Specific Activation of Endogenous Opioids," *Pain* 90 (2001): 205–215.

172 Benedetti's hypothesis is that: See the above-cited paper and M. Amanzio, and F. Benedetti, "Neuropharmacological Dissection of Placebo Analgesia: Expectation-Activated Opioid Systems Versus Conditioning-Activated Specific Subsystems," *Journal of Neuroscience* 19 (1999): 484–494.

173 There are other chemicals in the central nervous system: See paper by Donald D. Price and Dr. Lene Vase Soerensen, "Biological and Psychological Mechanisms of Placebo Analgesia," presented at the National Institutes of Health conference "The Science of the Placebo," November 19–21, 2000; and Hrobjartsson and Gotzsche, "Is the Placebo Powerless?"

174 Dr. Ted Kaptchuk, who studies placebos: Kaptchuk has a particularly interesting perspective given his background and training (which he describes in the PBS documentary cited in the note for page 168). After college in the United States, he studied medicine in China, and he is trained in traditional Asian techniques using acupuncture and herbs. Returning to the United States, he applied the traditional Western scientific methods of a placebo-controlled trial to assess the efficacy of acupuncture in ameliorating the pain of repetitive strain injury. Kaptchuk used "placebo acupuncture needles" that are visually identical to true needles and impart to the patients a similar sensation of being punctured but do not actually penetrate the skin. This controlled trial was being conducted in parallel with another randomized trial comparing amitriptyline versus placebo pill for the same condition. These two trials together created a four-arm trial that let Dr. Kaptchuk compare the placebo effect of an acupuncture sham device with the placebo effect of a sugar pill. Does a device or a more impressive ritual have more of a placebo effect than a pill

and a simpler ritual? How much does the dynamic of the patient-provider interaction affect treatment? This trial will help us understand what environmental cues trigger belief and expectancy and how this alters pain perception. Dr. Kaptchuk is also currently undertaking an experiment to see whether placebo effects can be delivered in a "dose-dependent" manner, similar to the action of many drugs.

175 In July 2002, *The New England Journal of Medicine*: J. B. Moseley, K. O'Malley, N. J. Petersen, T. J. Menke, B. A. Brody, D. H. Kuykendall, J. C. Hollingsworth, C. M. Ashton, and N. P. Wray, "A Controlled Trial of Arthroscopic Surgery for Osteoarthritis of the Knee," *The New England Journal of Medicine* 347 (2002): 81–88. Also see Margaret Talbot's article "Astonishing Medical Fact: Placebos Work!," in *The New York Times Magazine,* January 24, 2000.

177 Not long after the publication of the study on arthroscopic knee surgery: See Robert McGough's article "Barbecuing Your Back: How to Decide If New Heat Therapy Is Right for You," in *The Wall Street Journal,* February 11, 2003.

180 Now I look for ways to alleviate some of the fatigue: J. E. Groopman and L. M. Itri, "Chemotherapy-Induced Anemia in Adults: Incidence and Treatment," *Journal of the National Cancer Institute* 91 (1999): 1616–1634.

181 One clinical example of this involves asthma: See Anne Harrington, *The Placebo Effect: An Interdisciplinary Exploration* (Cambridge, Mass.: Harvard University Press, 1997); T. J. Luparello, N. Leist, C. H. Lourie, and P. Sweet, "The Interaction of Psychologic Stimuli and Pharmacologic Agents on Airway Reactivity in Asthmatic Subjects," *Psychosomatic Medicine* 32 (1970): 509–513; T. J. Luparello, H. A. Lyong, E. R. Bleecker, and E. R. McFadden, "Influences of Suggestion on Airway Reactivity in Asthmatic Subjects," *Psychosomatic Medicine* 30 (1968): 819–825.

185 It is likely that in certain settings, such as recovery from heart attack: See Bruce S. McEwen's article "Protective and Damaging

Effects of Stress Mediators," in *The New England Journal of Medicine* 338 (1998): 171–179; Harold Koenig, *Handbook of Religion and Health: A Century of Research Reviewed* (New York: Oxford University Press, 2000).

186 In August 2001, Dr. A. Jon Stoessl: Dr. Stoessl and the patient participating in this study were interviewed in the PBS documentary cited above in the note for page 167. The patient expressed amazement at the improvement in his motor function with placebo therapy. The videotape demonstrates this improvement as the patient is put through a series of maneuvers to test the range of motion of his limbs and his strength. See R. Fuente-Sernandez, T. J. Ruth, V. Sossi, M. Schulzer, D. B. Calne, and A. J. Stoessl, "Expectation and Dopamine Release: Mechanism of the Placebo Effect in Parkinson's Disease," *Science* 293 (2001): 1164–1166.

188 The "reward circuits" that Cohen is studying: P. Shizgal and A. Arvanitogiannis, "Gambling on Dopamine," *Science* 299 (2003): 1856–1858; C. D. Fiorillo, P. N. Tobler, and W. Schultz, "Discrete Coding of Reward Probability and Uncertainty by Dopamine Neurons," *Science* 299 (2003): 1898–1902.

189 Only very recently have studies begun to identify genetic differences: J. K. Zubieta, M. M. Heitzeg, Y. R. Smith, J. A. Bueller, K. Xu, Y. Xu, R. A. Koeppe, C. S. Stohler, and D. Goldman, "COMT $val^{158}met$ Genotype Affects μ-Opioid Neurotransmitter Responses to a Pain Stressor," *Science* 299 (2003): 1240–1243.

189 Environment also has a major influence on neural circuits: A recent review of neuroplasticity and its clinical implications can be found in H. Ikeda, B. Heinke, R. Ruscheweyh, and J. Sandkuhler, "Synaptic Plasticity in Spinal Lamina I Projection Neurons That Mediate Hyperalgesia," *Science* 299 (2003): 1237–1240.

190 Rather, events in our lives modify them: See Joseph LeDoux, *Synaptic Self: How Our Brains Become Who We Are* (New York: Viking, 2002).

CHAPTER 8: DECONSTRUCTING HOPE

191 Davidson is an experimental psychologist: Richard Davidson is
the William James and Vilas Research Professor of Psychology
and Psychiatry at the University of Wisconsin, Madison. For
more on his research interests and ongoing projects, consult his
website, http://psyphz.psych.wisc.edu.

192 "I was deeply affected by Norman O. Brown": See Norman O.
Brown, *Life Against Death: The Psychoanalytical Meaning of History*
(Middletown, Conn.: Wesleyan University Press, 1959) and
Love's Body (Berkeley: University of California Press, 1990).

192 "and by the writings of Herbert Marcuse": For example, see Herbert Marcuse, *Eros and Civilization: A Philosophical Inquiry into
Freud* (Boston: Beacon Press, 1974).

192 Even "emotion" is not uniformly applied by researchers in the
field: For example, Joseph LeDoux and Antonio Damasio used
different definitions of the terms "feeling" and "emotion" in their
work. See Joseph LeDoux, *The Emotional Brain: The Mysterious
Underpinnings of Emotional Life* (New York: Simon & Schuster,
1996) and Antonio Damasio, *Looking for Spinoza: Joy, Sorrow, and
the Feeling Brain* (New York: Harcourt, 2003).

193 "I understand hope as an emotion made up of two parts": This is
the working definition of hope that I cite in the Introduction.

194 "This is one of Antonio Damasio's most important contributions": See Antonio R. Damasio, *Descartes' Error: Emotion, Reason,
and the Human Brain* (New York: Avon Books, 1994).

195 Joseph LeDoux, professor of science at New York University:
For a lucid discussion of the biology of fear within the larger context of the complex subject of how emotions are generated and
how responses occur physiologically to these emotions, see
LeDoux, *The Emotional Brain*. Other overviews of the biology of
fear and anxiety are found in "Our Bodies, Our Fears," *Newsweek,*
February 23, 2003, and "Understanding Anxiety," *Time,* June 10,
2002.

196 The case of a young woman S.M.: R. Adolphs, D. Tranel, H. Damasio, and A. Damasio, "Impaired Recognition of Emotion in Facial Expressions Following Bilateral Damage to the Human Amygdala," *Nature* 372 (1994): 669–672; S. B. Hermann, L. Stefanacci, L. R. Squire, R. Adolphs, D. Tranel, H. Damasio, and A. Damasio, "Recognizing Facial Emotion," *Nature* 379 (1996): 497; R. Adolphs, D. Tranel, and A. R. Damasio, "The Human Amygdala in Social Judgment," *Nature* 393 (1998): 470–474.

202 One of the intriguing but still very controversial findings: See Joseph LeDoux, *Synaptic Self: How Our Brains Become Who We Are* (New York: Viking, 2002); Esther M. Sternberg and Philip W. Gold, "The Mind-Body Interaction in Disease," *Scientific American,* Special Edition: The Hidden Mind, 12 (2002); Esther Sternberg, *The Balance Within: The Science Connecting Health and Emotions* (New York: W. H. Freeman & Co., 2000).

202 An alternative hypothesis, based on studying identical twins: See M. W. Gilbertson, M. E. Shenton, A. Ciszewski, K. Kasai, N. B. Lasko, S. P. Orr, and R. K. Pitman, "Smaller Hippocampal Volume Predicts Pathologic Vulnerability to Psychological Trauma," *Nature Neuroscience* 5 (2002): 1242–1247.

204 To quantitate the women's handling of stress, Davidson and his coworkers assessed changes in the amount of cortisol: A detailed discussion of the hypothalamic-pituitary-adrenal axis that regulates cortisol release in the body is found in Sternberg and Gold, "The Mind-Body Interaction in Disease," and Sternberg, *The Balance Within*.

Index

Beckwith, George, 101
belief:
  hope and, 167, 170, 173, 176–77, 185
  in placebo effect, 167, 169, 172, 173,
    174, 176–78, 181–85, 187–88
Benedetti, Fabrizio, 170–74, 177
Beth Israel Deaconess Medical Center, 151
Bible:
  on illness as punishment for sin, 20
  Noah story in, 123
Billings Hospital, 141
biomedical studies, available through
  National Library of Medicine, 165
Bird, Larry, back surgery and
  rehabilitation of, 151–52, 153–54
blood transfusions, anemia and, 117
body, definition of, 167
body brace, 148
body-mind connection:
  and fatigue, 179–80
  vicious cycle in, 94–95, 119–20,
    155–57, 178, 179–80, 185–86, 199
bone marrow transplants, 98
Boston Celtics, 152, 154
Boston Marathon, 147
bowel, ischemic, 111
brain:
  discrete function centers of, 201
  dopamine production in, 187–89, 200
  and drug addiction, 165, 166, 188
  hope and chemistry of, xvi, 156–57,
    159, 160, 166, 170, 172, 173, 174,
    177, 184–85, 186–89, 190, 198,
    199–207
  hope or despair shaped by visceral input
    to, 94–95, 119–20, 155–57,
    159–60, 178–81, 185–86, 199
  mind as manifestation of, 166–67
  neuronal interlacing circuits of, 201
  and pain perception, 159–60, 169–74
  plasticity of, 189–90, 200
  reward and goal-seeking circuits of, see
    reward and goal-seeking circuits
  see also specific parts of brain
brain research, in early stages, 166, 206
brain scans:
  of discrete function regions, 201

in Parkinson's placebo experiments,
  187–88
breast cancer, 4–24, 25–27, 44, 82,
  121–46, 211
  biopsy in, 7, 9, 121–22, 124
  chemotherapy for, 11–12, 14, 19–20,
    43, 122, 124–25, 127, 128, 129,
    130–35
  inflammatory, 80–81
  metastatic, 4–24, 42–43, 121–46
  palliative care for, 125–26, 128–29,
    131–34
  physical examination for, 6
  risk factor for, 5–6
  surgical options in, 7, 8–10, 14, 17,
    122
  tumors in, 4, 6, 7, 9, 11, 71, 122,
    123–24, 126, 130, 132–34
bronchi, in asthma, 181–85
bronchoalveolar carcinoma, 66, 67
bronchodilators, 182–83, 185

calcium, in Urbach-Wiethe disease, 196
California, University of, at Los Angeles,
  medical center of, Groopman's
  fellowship at, 28, 29, 35–36, 42–44,
  50, 56, 116
cancer:
  cure vs. palliation in, 32
  deaths from, in 1980s, 82
  HIV and, 82
  remission vs. cure in, 35, 39, 125
  see also oncologists, oncology; tumors;
    specific diseases
carcinogens, xiv
  in stomach cancer, 65–66
Cassell, Eric, 132
CAT scans, 32, 35
cells, on and off:
  opiates and, 170
  pain and, 169–70, 172
central nervous system:
  autonomic nervous system and, 181
  opiate effect on, 170
  in pain perception and emotions,
    169–70, 179, 189

cure:
  in cancer, 32, 35, 73, 77–78, 79–80,
    118, 134
  "miracle," 58–81
  palliation vs., 32
  remission vs., 35, 39, 125

Damasio, Antinio, 194–95, 196–98, 210
D'Angelo, Catherine, 80–81
Davidson, Richard, 191–206, 209
  work focus of, 191, 192, 203–6
death, dying:
  body-mind connection and impending
    sense of, 88–89, 92, 94–95, 96,
    105, 109, 114, 119–20, 178,
    179–80, 185–86
  from cancer and AIDS in 1980s,
    82–83
  five stages of, 142–44
  Groopman's father's attitude toward,
    145, 146
  as patient failure in mind-body
    connection doctrine, 163
  and patients' end-of-life directives, 54,
    99, 108, 126–27
  serenity in face of, 127, 135–36,
    137–38, 144–46
debridement, in knee surgery, 175–76
declarative memory, 201–2
defoliants, 86
depression, clinical, 164
diarrhea, chemotherapy and, 31, 34
discectomy, 147–48
Doctoring (Cassell), 132
doctors:
  alternative healers vs., 93
  in conveying of poor prognoses, 32, 51,
    75
  as "cowboys," 67
  emotions and anxieties
    compartmentalized by, 44, 145–46
  fallibility of, 51, 209
  false hopes as concern of, 61, 81, 144,
    210
  false hopes raised by, 31, 33, 34, 35, 37,
    39–40, 41–42, 43–44, 51–52, 53,
    144, 208

friendships between patients and, 132,
    135–41
  helping patients find hope, 53–56,
    60–61, 79, 81, 84, 144, 209–12
  joke about God, 129–30
  nonmedical lessons for, 121
  patients' expectations often pinned to
    attitudes of, 51, 174, 185, 193
  patronizing and writing off patients by,
    41–42, 51–52, 53, 72–73, 75, 79
  proper place of, at time of diagnosis,
    79, 81, 84, 210
  training of, see medical training
Dolan, Deirdre, 115–17, 120
dopamine:
  genes and metabolism of, 189
  in goal-seeking and reward circuits of
    brain, 188–200
  Parkinson's disease and, 187–88
drug addiction:
  biological basis of, 165, 166, 188
  "reward circuits" in, 188
drugs, illegal, 88

emotions:
  amygdala damage and, 196–98
  cognition, deliberative thinking and,
    193–95, 196–99, 210
  doctors and compartmentalizing of, 44,
    145–46
  James on neural input and, 178
  negative, see negative emotions
  positive, see positive emotions
end-of-life directives, 54, 99, 108, 126–27
endorphins, 190
  blocking of, 173, 179
  genes and release of, 189
  hope and release of, xvi, 170, 172, 173,
    174, 177, 179
enkephalins, 190
  blocking of, 173, 179
  genes and release of, 189
  hope and release of, xvi, 170, 172, 173,
    174, 177, 179
erythropoietin, 117, 125, 180
esophagus, stomach cancer and, 69–70
estrogen, breast cancer and, 5–6, 80–81

patients (*cont'd*)
  doctors' role in finding hope for,
    53–56, 60–61, 79, 81, 84, 144,
    209–12
  empowerment of, 92–93
  expectations of, 51, 174, 185, 193
  friendships between doctors and, 132,
    135–41
  in helping other patients find hope, 78,
    116–17, 120, 210
  hope and choice as rights of, xiii,
    52–53, 75, 78, 79, 81, 84, 126–27,
    144, 210
  impending sense of death in, 88–89,
    92, 94–95, 96, 105, 109, 114,
    119–20, 178, 179–80, 185–86
  inaccurate information given to, 31, 33,
    34, 35, 37, 39, 41–42, 43–44, 51,
    52–53, 75, 144
  Kübler-Ross and study of, 141–44
  labelling of, as doctors' distancing
    mechanism, 26–27
  medical students' interactions with, 6,
    15, 16, 21, 22
  stark prognoses given to, 42–43, 50,
    51, 52–53
Paulsen, Eric, 58–59, 60, 67, 71, 81
peritonitis, 110, 112
phrenology, 200
physical therapists, physical therapy:
  at Baptist rehabilitation program,
    157–59
  for elderly, 149
  passive, 149
Pierson, James, 64
pituitary gland, 196
placebo effect, placebos, 167–78, 181–85,
    186–90
  authority figures and, 171, 174, 177
  autonomic nervous system and,
    181–85
  belief and expectation in, 167, 169,
    172, 173, 174, 176–78, 181–85,
    187–88
  brain cortex and, 186–88
  changing scientific views of, 168–69
  definitions of, 168
  environmental cues in, 174, 177, 184

prevention of, 173
  "reward circuits" and, 188
  spectrum of response to, 189–90
placebo effect, placebos, experiments
    with, 167–78, 181–85, 186–90
  asthma in, 181–85
  Benedetti and, 170–74, 177
  1950s clinical trials, 168
  pain and, 169, 170–78, 189
  Parkinson's disease in, 186–88
  sham arthroscopic knee surgery in,
    175–77
  sham back surgery in, 177–78
Poles, stomach cancer and, 65
positive emotions:
  biology of, 191–207
  connections between negative emotions
    and, 199–207
  new studies on, 164
  pain ameliorated by recalling of, 159
  popular literature on health and,
    xv–xvi, 162–63
  see also body-mind connection; mind-
    body connection
positron emission tomography (PET) scan,
    187–88
post-traumatic stress disorder, 202
prayer, 77, 129, 211
prefrontal cortex, reward circuits and,
    188, 199–200
preserved foods, 65–66
proteins, xv
Psalm, Twenty-third, 76, 78–79, 211

radiation therapy, 48–49, 54–56, 58–59
  for breast cancer, 80, 122
  for cholangiocarcinoma, 48–49, 50,
    54–56
  for lung cancer, 80
  lymphoma and, 83–84, 90, 91, 92, 98,
    106–7, 118, 186
  partial response to, 68
  side effects of, 50, 56, 60, 67, 70, 72,
    80
  for stomach cancer, 58–59, 60, 67, 68,
    70, 71, 72–73, 74, 77
  tumor lysis and, 86, 106–7

JEROME GROOPMAN, M.D., holds the Dina and Raphael Recanati Chair of Medicine at the Harvard Medical School and is the chief of experimental medicine at the Beth Israel Deaconess Medical Center in Boston. His research has focused on the basic mechanisms of blood disease, cancer, and AIDS. He is a staff writer in medicine and biology for *The New Yorker* and is the author of two popular books, *The Measure of Our Days* and *Second Opinions,* which were the inspiration for the television series *Gideon's Crossing.* In 2000 he was elected to the Institute of Medicine of the National Academy of Sciences. He lives with his wife and three children in Brookline, Massachusetts.